MIT AND THE BIRTH OF DIGITAL SIGNAL PROCESSING

Volume 4 in the "Scientist and Science" series

Enders Anthony Robinson

Professor Emeritus in the
Maurice Ewing and J. Lamar Worzel Chair
Columbia University in the City of New York

Goose Pond Press

Available from Amazon.com and other retail outlets.

This book is dedicated to
MIT President James Rhyne Killian

In 1951, when funding avenues failed, he put his presidential reserves on the line to open the way for the first signal processing and deconvolution ever done on a digital computer.

Copyright © 2015

by

Enders Anthony Robinson

All Rights Reserved Worldwide

Goose Pond Press

Ralph Waldo Emerson told Henry David Thoreau that Goose Pond should be called The Droplet or God's Pond. It was significant to many of Concord's leading literary figures, all of whom walked there often.

James Rhyne Killian, Jr.

July 24, 1904 – January 29, 1988

The 10th president of the Massachusetts Institute of Technology

from 1948 to 1959

Contents

Preface 7

Chapter 1. Norbert Wiener and digital signal processing 9
    IEEE History 9
    Seismic reflection method 12
    Detection of reflections, vintage 1950 16
    MIT in 1950 17
    Letter from a MIT student 49 years later 18
    Eight seismic field records 18
    MIT Mathematics Department in 1951 20
    Origin of deconvolution 24

Chapter 2. MIT President James R. Killian 28
    MIT Department of Geology and Geophysics 28
    Christmas present in 1951 35
    History of Whirlwind 37
    Deconvolution implemented on Whirlwind 1952 38
    The MIT meeting of August 1952 42
    Raytheon in 1952 45

Chapter 3. Geophysical Analysis Group 53
    Geophysical Analysis Group in 1953 53
    Geophysical Analysis Group in 1954 57
    Geophysical Analysis Group in 1954-1957 62
    Geophysical Analysis Group 1954 and Beyond 67
    Official citation for asteroid Svenders: 71
    The growth of digital signal processing 72

Chapter 4. MIT Reports 74

Table of Contents

MIT Reports on Research 74
Oil from mathematics (MIT, 1953) 75
Making Electrons Count 77
Chapter 5. Raytheon Computer Services 79
   IEEE Computer Pioneer Award 79
   Discussion of machine solution of seismogram analysis problem 82
Chapter 6. MIT PhD thesis 1954 90
   Letter from Professor Herman Wold 90
   Chapter VI of MIT PhD thesis 1954 91
   Mike Perz and Gary Margrave 108
Chapter 7. Andre Kolmogorov 111
   The wavelet 114
   Canonical representation 116
   All pass z-transform 118
   Wold decomposition 119
   Extremal properties 120
   Linear prediction 122
   References 128
Chapter 8. Digicon 130
   Texas Instruments 130
   The Origins of Digicon 132
   Letter from Pat Poe 1965 136
   A history written by the geophysical company CGG 136
Chapter 9. A Historical Perspective of Spectrum Estimation 139
   John Tukey, Claude Shannon, Hendrik Bode 139
   Spectral estimation 142
   Taylor series 145

Daniel Bernoulli solution of the wave equation 147

Jean Baptiste Joseph de Fourier and the sinusoidal spectral theory 149

Sturm-Liouville spectral theory of differential equations.................. 152

Schrodinger spectral theory of the atom.. 157

The von Neumann spectral representation theorem 162

Einstein-Wiener theory of Brownian motion..................................... 168

Yule autoregressive spectrum estimation method............................ 173

Wiener's generalized harmonic analysis.. 177

Reconstruction of the two spectral theories 183

Wiener-Levinson prediction theory 185

Tukey empirical spectral analysis... 192

Cooley-Tukey fast Fourier transform 194

Predictive deconvolution (aka blind deconvolution)........................ 195

Statistical theory of spectrum estimation... 201

Engineering use of spectral estimation.. 203

Acknowledgement 203

References 204

Preface

In the late 1950s, the first transistorized digital computers were manufactured. As a result, the decade of the 1960s saw the conversion of the oil exploration industry from analog filtering to digital signal processing. Digital signal processing and deconvolution made it possible to explore successfully in all areas of the earth including the areas under the seas.

Deconvolution requires massive parallel processing. The computer industry accommodated and developed special-purpose digital signal processing (DSP) computers, which at the time were called array processors (also known as vector processors). As time went on, these DSP computers were replaced by microchips, as were all digital computers.

In 1986 the National Academy of Engineering (NAE) of the United States of America elected Enders Robinson as member with the citation:

> "For pioneering contributions that have led to the evolution of seismic processing from hand digitization of the 1950s to today's custom deconvolution chip"

The hand digitization of the 1950s was the forerunner of what was to come. Practically all communications today are done with digital data. The telephone is digital, the television is digital, the cinema is digital, the music is digital, the camera is digital. Now in the second decade of the twenty-first century, DSP chips are everywhere.

The deconvolution chip is the prototype digital signal processing (DSP) chip. Usually the NAE credits new members for such things as "innovative research" or "revolutionary discoveries." Certainly such chips are vital, but why would the NAE mention a chip and not mention seemingly more important mathematical research?

The National Academy foresaw what many others did not foresee. In 1986 the Academy realized that the era of mathematical numeral analysis was losing its position at the forefront of computing. Instead the era of communication was at the beginning of its meteoric rise to ascendency.

Communication requires a massive use of digital signal processing. Today in 2014, digital signal processing is a powerful technology used everywhere. This technology is universally possible because of inexpensive and tiny DSP (Digital Signal Processing) chips. DSP chips are an integral part of audio and video systems, television, telephones, cameras, radar, sonar, transmission fiber cable, automobiles, airplanes, ships, space vehicles, programmable heart pacemakers, other medical devices, and all types of automation in manufacturing and logistics.

Digital Signal Processing (DSP) chips are used to satisfy the insatiable needs for communication among people and for the regulation and control of machines and devices. DSP chips take real-world signals like position, voice, audio, video, pressure, temperature, and medical readings to process them digitally in order to produce the desired results. Every time you use a mobile phone, you are using a DSP chip. Every time you drive an automobile, you are using several dozen DSP chips. The new medical devices that monitor you heath all use DSP chips.

The birth of digital signal processing took place in the MIT Mathematics Department from 1950 to 1952 under the guidance of Prof. Norbert Wiener. The DSP chip uses the methods of digital signal processing that were run for the first time on the Whirlwind computer at MIT in 1952. This book takes you on a journey showing how this development evolved and led to the DSP chip.

This development took place because MIT President James R. Killian challenged the conventional analog thinking of the time. He stepped in and put up his presidential funds to support the DSP project that otherwise would have perished.

Chapter 1. Norbert Wiener and digital signal processing

IEEE History

SIGNAL PROCESSING

THE EMERGENCE OF A DISCIPLINE
1948 TO 1998

Frederik Nebeker

IEEE History Center
Rutgers - The State University of New Jersey

1998

In 1998, Frederik Nebeker wrote the book *SIGNAL PROCESSING, The Emergence of a Discipline, 1948–1998*. It gives an account of one of the greatest technological innovations of the twentieth century—signal processing. In documenting the evolution of an engineering discipline, the book is marvelously revealing of the way science is done. The book is a part of an undertaking of the Institute of Electrical and Electronic Engineers (IEEE) to document the history of engineering.

The rest of this section consists of excerpts from Nebeker's book.

The story is told chronologically, beginning with the year 1948, though many earlier achievements are mentioned. .... Though no one doubts that the engineers and scientists of Bell Telephone Laboratories and the Massachusetts Institute of Technology played a huge role in the development of the field, this account probably gives too much space to their work. .... There are many reasons to study the history of one's field. Knowing how a field developed gives one a deeper understanding of the present state-of-the-art. The scientific, economic, and social forces that influence technological developments are clearer when viewed over time. And one can gain a greater sense of purpose from seeing one's work as part of a long tradition of endeavor that has elicited the energy and ingenuity of countless admirable people. This modest first attempt to survey the development of a major area of engineering will certainly be improved upon. One of the functions of the IEEE History Center is to stimulate engineers to research and write the history of their technical areas, a function it is hoped this monograph will serve.

Chapter 2   Birth Certificate of the Information Age: The Annus Mirabilis 1948

ENDERS ROBINSON: I came back to MIT in the fall of 1950, and I was in the Mathematics Department as a graduate research assistant. I was working under Norbert Wiener to find applications for his time-series analysis. As you know, Norbert Wiener was an eminent mathematician at MIT, and he worked in generalized harmonic analysis back in 1930. He always felt that people didn't realize that was probably his most important contribution. But during World War II, he worked on prediction theories for anti-aircraft fire control. He developed a theory—classified at the time—but published by MIT Press in 1949. We used the book when I was an undergraduate student there, so I was familiar with it. By 1950, we were ready to apply it. Meanwhile, Wiener had written another book called *Cybernetics* which was first published about 1948. That book was an instant success; he was the one that introduced the word cybernetics.

So Wiener was a celebrity by 1950, and my research assistantship depended on finding applications of his work. Another MIT mathematics professor, George Wadsworth, my graduate advisor, was working in weather prediction which he originally started during World War II; Wiener's classified book had come out, but it was too difficult mathematically to be applied as such. So Wadsworth asked Norman Levinson, who was another eminent mathematician at MIT, to take Wiener's book and to simplify it into numerical algorithms that he, Wadsworth, could use. Levinson published these papers in the *Journal of Mathematics and Physics* in 1947. They were added as appendices to Wiener's 1949 book, so Levinson's algorithms with Wiener's theory became the way to do this type of thing. We could then apply it to geophysics because seismic records are essentially noisy records.[1]

> **ENDERS ROBINSON:** So Professor Hurley obtained eight seismic records, like this one, so that I actually had the data in the fall of 1950.... The first step was for me to hand-digitize these eight records by putting a T-square down with a scale and reading off the traces point-by-point, putting them in numerical form.
> **INTERVIEWER:** You probably had a pretty modest sample rate!
> **ROBINSON:** Yes ... it took a couple of days to do a record. And then you'd have to check to make sure.[1]
>
> ---
>
> **ENDERS ROBINSON:** To get back to 1951, we had a method that worked, which was called deconvolution. The next question was how to compute it. It took all summer to deconvolve a few of these traces with Virginia Woodward [working with a Marchant desk calculator]. At that time MIT had a computer they called Whirlwind.... Like the ENIAC, it took whole rooms, a whole building, the Barta Building at MIT.... So in the spring of 1952 I went to Whirlwind with Howard Briscoe and put deconvolution on Whirlwind.[2]

> **ENDERS ROBINSON:** The water layer itself is reverberating like a drumhead and that hid the signals coming from the depths. Deconvolution removed those reverberations.
> By the late 1950s there were lots of areas they wanted to explore offshore, like the Gulf of Mexico and the Persian Gulf, and in Venezuela they had Lake Maracaibo. They realized that the only way they could get rid of water reverberations was by deconvolution. The only way they could deconvolve was by signal processing, because analog cannot do it. So they were sort of forced into digital that way.
> ... The analog methods they used would be electrical band-pass filtering, high-pass filtering, and low-pass filtering. They could also adjust the traces in time: move one trace with respect to the other. In other words, they might think, "Well, the signal's coming in at 40 hertz, so we will band-pass it close to 40 hertz to find the signal." That would be analog processing because they did it by an electric circuit.
> **INTERVIEWER:** And that was supposed to strip out the reverberations?
> **ROBINSON:** They could use band-pass filtering if the reverberations were at a different frequency than the deep reflections. That was the idea, but it didn't work because all these signals overlapped in frequency. The companies decided they had to get rid of those reverberations, which they could do through deconvolution. That meant going digital, so they started spending the money. That was the advantage of the oil industry, they had money. The other thing is that they had a history of spending money on computers through their refining operations. They had close contacts with IBM and other computer companies.[1]

On Pages 51 and 52, Frederik Nebeker writes:

Analysis of seismic data, like analysis of radar data, stimulated the development both of computing technology and signal-processing techniques. Texas Instruments, originally a manufacturer of seismic instruments used in petroleum exploration,[59] began in 1956 to design a digital computer for processing seismic data.[60] Signal-processing technique was advanced by, among others, Enders Robinson. In the early 1950s he showed how to derive the desired reflection signals from seismic data, carrying out one-dimensional deconvolution.[61] The digitization and calculations were carried out by hand on a desk calculator; the deconvolution of 32 traces, each one 600 to 800 readings, took the entire summer of 1951.[62] In the spring of 1952 Robinson and Howard Briscoe programmed the MIT Whirlwind digital computer to do the numerical filtering at high speed.[63] A group at the Raytheon company contracted with MIT to do programming and computation tasks relating to the analysis of seismograms, and in March 1954 Raytheon offered to the industry at large what must have been the first commercial digital-signal-processing service.[64]

Seismic reflection method

We want to fill in the details of Nebeker's story given above. The birth of digital signal processing took place in the MIT Mathematics Department in 1950–1952 under the guidance of Prof. Norbert Wiener and with the support of Prof. George Wadsworth and Prof. Norman Levinson. Also Professor Paul Samuelson and Prof. Robert Solow of the Economics Department were actively involved. All of the other projects making use of Wiener's work were analog. Our project was the only one that was digital and, in due course, it led to the Digital Signal Processing (DSP) chip that is universally present in today's world of communications.

Certain animals have senses that humans do not possess that permit them to do sophisticated methods of remote sensing. Several species of fish, sharks, and rays have the capacity to sense changes in electric fields in their immediate vicinity. Some fish passively sense nearby electric fields, whereas other fish actively transmit their own weak electric signals and sense the pattern of returned electric signals over the surface of their bodies. Among mammals, the platypus has the most acute sense of electroreception. Some birds and insects such as bees have the ability to detect fluctuations in magnetic fields. Although this sense is not a well understood, it is essential to the navigational abilities of migratory birds. Spiders can detect mechanical strain in the exoskeleton, providing information on force and vibrations. Bats and cetaceans are noted for their use of echolocation—the ability to determine orientation to other objects through interpretation of reflected sound. Echolocation guides whales and dolphins through the darkness of the deep and helps them to identify prey.

Remote sensing is the acquisition of information of an object by the use of sensing devices that are not in physical or intimate contact with the object. Remote sensing is dependent upon the use of signals that allow communication between the observer and the observed. The most important type of signal is the traveling wave, whether mechanical waves as in sound or electromagnetic waves as in light. There are two kinds of remote sensing: passive and active. In passive remote sensing, the observer waits for signals emitted by unknown and/or unreachable objects. In active remote sensing, the observer emits signals and then records the resulting signals that are reflected, refracted, or scattered by the object.

Radar and sonar are premier methods of echolocation. Radar uses electromagnetic signals for active remote sensing. The radar antenna sends out microwaves. Radar receivers are usually, but not always, in the same location as the transmitter. The reflected radar signals captured by the receiving antenna are usually very weak. The original radar sets were analog. They strengthened the received waves by amplifiers and displayed them on an analog screen.

The earth is almost a sphere, but the centrifugal force of its rotation causes a bulging at the equator and a slight flattening at the poles. If we could see the earth in cross section, we would find a sharp division between the core, or central part, and the mantle, or outer part. We would also find that the outer surface of the mantle has a very shallow skin layer of different composition, known as the crust. But we cannot see deep within the earth. How did we obtain such knowledge? Earthquakes generate seismic waves. The seismic waves serve as the signals required for remote sensing. Because we must wait for earthquakes to occur, earthquake seismology is passive. Petroleum and other mineral resources are located in the upper layers of the crust. In seismic exploration, source signals are transmitted into the ground, and the reflected seismic waves are recorded. Exploration seismology represents active remote sensing.

The key aspect of remote sensing resides in the acquisition of observational data in the form of signals. The premier sensing device is the eye, whether of an animal or a human. The eye receives a huge number of electromagnetic signals (light) coming from all directions all the time. All of these signals are processed by the brain. The result is vision, which lets a person or animal see. Sight is so natural so we take the signal processing in the brain for granted. But think of a bat, which is blind. It sees through its ears. By echolocation it can find a hole the thickness of a man's thumb and can squeeze through it. Sound waves have huge wavelengths as compared to light waves, and yet by signal processing the bat's brain processes the reflected sound to the extent that it can achieve enough signal resolution to form a meaningful image.

The general principle of seismic reflection is to send elastic waves (using an energy source such as dynamite explosion or vibroseis) into the earth, where each layer within the earth reflects a portion of the wave's energy back and allows the rest to refract through. These reflected energy waves are recorded over a predetermined time period (called the record length) by receivers placed either on or close to the surface of the earth. On land, the typical receiver used is a small, portable instrument known as a geophone, which converts ground motion into a recordable signal. In water, hydrophones are used, which convert pressure change into a recordable signal. A record of the data from one seismic channel is known as a "trace." The traces from all of the receivers are recorded. The shot location is moved along and the process is repeated.

Remote sensing is a basic element in the observational sciences, and in particular remote sensing is fundamental to reflection seismology. The explorers of old traversed the seven seas to find new lands and then sailed home with their treasures. Geophysicists cannot travel through thousands of feet of solid rock to find new horizons. Instead, they must seek treasure indirectly by sending signals into the ground and interpreting their echoes. Exploration seismology represents a case of remote sensing. In the early days, the geophysicists made interpretations directly from the observations acquired. A seismic wave is a mechanical perturbation that travels in the earth at a velocity that depends upon the medium (rock) in which the wave is travelling. When a seismic wave travelling through the earth encounters an interface between two rock layers with different acoustic impedances, some of the wave energy will reflect off the interface and some will refract through the interface.

The seismic reflection technique requires the generation of seismic waves. The waves travel from the source, reflect off an interface, and are detected by an array of receivers (or geophones) at the surface. To an uninitiated observer the received signals can be misleading and inconclusive. With a very large degree of skill and know-how, the seismic interpreter had to pick out the arrival times of the reflected pulses and plot them in order to obtain a map of the underlying structure of the earth. The thrill of seeing this picture gradually emerge from a mass of seemingly non-interpretable data was the reward. The results obtained may be sensitive to relatively small errors in the observations of the reflected pulses. Great care must be taken when interpreting the results of a reflection seismic survey.

Seismic exploration is an echolocation scheme. It does exactly what bats do naturally. The bats interpret returned sound signals to detect the echoes needed to guide them in flight. A seismic reflection in rock is analogous to an echo of sound in air. For example, a geophysicist activates a source that emits seismic waves into the earth. The seismic waves travel outward. A seismic wave continues its path until it strikes an interface where it is reflected. When this reflected seismic wave returns to the surface of the earth, it is picked up by a detector (geophone) and is recorded as a trace (wiggly line) on a seismogram. Many detectors are used a various surface points. There is a trace for each detector.

Exploration seismology has the distinction of being the first observational science that made extensive use of the digital signal processing. In so doing, the geophysicist produces accurate and reliable digital images of the underlying structure of the earth. Today, the recorded seismic signals are subjected to significant amounts of digital signal processing in order to produce accurate images of the subsurface. Large amounts of computer processing are performed on supercomputers or computer clusters.

Geophysical methods are used extensively in the exploration for oil and gas. These techniques exploit the diagnostic capabilities of practically all the observable physical properties of the earth. To locate the deposits of oil, the geophysicist must delineate formations that lie beneath the surface. The most successful method is the seismic method. This method exploits the elastic properties of rocks. A man-made energy source such as an explosion is located on or close to the surface. This source excites elastic waves in the rocks (seismic waves) which in various forms propagate through the subsurface rock layers. Some of these waves penetrate to depths of thousands of feet where they are reflected from various interfaces. A portion of these reflected waves return to the surface where they are recorded in the form of seismic traces on a seismogram. The seismic data can then be analyzed and interpreted so as to yield valuable information as to the subsurface

This need for the high-resolution mapping of the three-dimensional underground structure of the earth falls upon the geophysicist. His task is made possible by digital signal processing.

Detection of reflections, vintage 1950

In 1950, everything was analog. The geophysicists recorded the retuned seismic signals as tracings on rolls of photographic paper. They looked at the traces to detect the reflections (i.e., seismic echoes) needed to guide them in mapping the subsurface of the earth. In the terms of the trade, the geophysicists would "eyeball" the traces for the "primary reflections." In other words, the recognition was done visually by a person called a seismic interpreter. The talent to distinguish the reflected signals from all other signals was acquired by practice and by knowledge of the various wave paths and properties.

What was the problem in 1950? The problem of first importance was the recognition of reflections on the seismic records. In many areas of the world, in fact in most areas, the geophysicists could not see the reflections. The reflections were too weak to poke their heads above all of the other seismic activity recorded on the trace. Such no-reflection records were virtually useless for geological interpretation.

The problem becomes one of recognizing the true primary reflections from the complexity of unwanted signals with which it is associated. This was the major problem in 1950. In only relatively few areas the reflections could be readily recognized. In most areas the complexity of impulses masked the reflections to such an extent that the desired reflections were unrecognizable.

MIT in 1950

I graduated in June 1950 from MIT, and then I went on active duty as a second lieutenant in the army at the outbreak of the Korean War. I spent that summer at Aberdeen Proving Grounds attending the Army Ordnance School. Aberdeen was a special place. The ENIAC was there, but locked away because its work was classified as secret. It was almost as if the ENIAC were enclosed within impenetrable walls. To be so close, and yet so far, from one of the greatest inventions of all time was mesmerizing. The ENIAC was a forbidden attraction, not unlike the ones which fascinated the Odysseus of mythology. In my spare time I learned as much as possible about the ENIAC. The ENIAC, the first high speed, electronic stored program digital computer, was the harbinger of the new world into which mankind would be thrust.

I graduated from the Army Ordnance School in September 1950, ranked second in the class in the subject of ordnance management (logistics). I was assigned to an early ready reserve unit at Watertown Arsenal in Massachusetts. The Army wanted MIT graduates at its manufacturing arsenals. Watertown manufactured essentially all of the antiaircraft guns and most of the heavy artillery used by the Army in World War II.

I started graduate school at MIT in September of 1950. I had a quarter-time teaching assistantship and also a quarter-time research assistantship in the mathematics department. For the teaching assistantship, I taught one section of freshman calculus.

Letter from a MIT student 49 years later

12 Jan 1999

Dear Prof. Robinson:

I was your student in a freshman math section at MIT in the Fall Term of 1950. I went on to a PhD. in aero from Princeton.

I followed your work in wavelets over the years and I was surprised that recent publications in that field frequently fail to cite your work. In particular I was re-reading the *Proceedings of the IEEE* April 1996 this week and that got me started. I am going to respond to an invitation by the Editorial Board of IEEE Proceedings to comment and I was going to say something about the provincial nature of the sources referenced. The reviews make it look like nothing of consequence, (with maybe one or two exceptions e.g. Gabor) published in the field between, say, 1930 and 1980 and that is not the case. I wanted to check with you before I comment. I can't claim perfection in checking all the references but I am pretty certain I didn't miss any citations to Robinson. The fact that many of the references are in geophysics doesn't excuse the authors. Geophysics is the field where the action started.

I was the Deputy for Technology in the Office of the Secretary of the Air Force for nearly seven years, until June 1991, and I was responsible for the entire USAF Science and Technology program. We had a problem then, (which I believe still remains,) in the resistance of the Lab researchers to looking up prior work especially across disciplines.

I look forward to hearing from you and I will send you a copy of my comments before I send anything on to IEEE or others on this subject.

Sincerely yours,

Bernard Paiewonsky

Eight seismic field records

Prof. Philip M. Morse of the MIT Physics Department and Prof. George P. Wadsworth of the MIT Mathematics Department were prominent among the small group of pioneers who in the 1940s originated the disciple of operations research. In 1947 Prof. Wadsworth was awarded a certificate of appreciation from the US government for his wartime

Chapter 1. Norbert Wiener and digital signal processing

research in meteorology, using statistical methods in weather forecasting for the European Theater of Operations.

For my quarter-time research assistantship in 1950, Professor Wadsworth gave me eight paper oscillographic multi-trace seismograms recorded by the Magnolia Petroleum Company. Magnolia was an exploration subsidiary of the Socony-Vacuum Oil Company in New York. (Socony is an acronym for **S**tandard **O**il **C**ompany **O**f **N**ew **Y**ork, now a part of Exxon Mobil.) My assignment was to find a method to detect the missing reflections.

The Mathematics Department expected me to attack the problem in the conventional theoretical way involving the mathematical solution of partial differential equations. This approach represents forward modeling. It goes from a theoretical model for the earth to theoretical seismic data. The differential equation models the earth. The model yields theoretical seismic data. The theoretical seismic data is compared to actual seismic data in order to test the efficacy of the model. However, they left me alone to do what I wanted to do.

I started with the actual seismic data. This approach represents inverse modeling. It goes from the actual seismic data to an estimated model for the earth. A numerical method had to be devised to find the primary reflections, which represent the interfaces in a layered earth. When a well is drilled the actual layering is revealed. The accuracy of the estimated model is then exposed for all to see. The power of this approach rests in the ability to find all of the primary reflections, even those that are hidden from view because of the signal-generated noise. I needed to find a numerical way to detect the unseen reflections.

As a first step, I sat down with pencil and straight-edge, and diligently converted each analog seismic trace on the paper records into digital form, a list of numbers. The second step would be to come up with a numerical method to detect the hidden reflections. The third step would be to do the calculations on a digital computer such as the ENIAC. I had written down the digitized seismic data in columns. This represented the hand-digitization. I was ready for the second step, namely to devise a digital method of analysis. The first thing I did was to compute the spectra and the cross-spectra of seismic traces. However this work gave no indications as to how the hidden reflections could be detected.

MIT Mathematics Department in 1951

Norbert Wiener (1894–1964) was a Professor of Mathematics at MIT. He did early work in stochastic and noise processes. His contributions were relevant to electronic communication and control systems. He originated the subject of cybernetics, which is a formalization of the notion of feedback, with implications in engineering, computer science, biology, neuroscience, and the organization of systems.

During World War II, Professor Norbert Wiener had occupied himself with the problem of the prediction of aircraft trajectories for the aiming of antiaircraft guns. His results were in his book *Extrapolation, Interpolation and Smoothing of Stationary Time Series*, published by the MIT Press in 1949. I was familiar with that book.

Professor Norbert Wiener was overseeing the seismic research. He had a great interest in the applications of theory to practice. In the 1951 spring semester, Professor Wiener went on leave to be in Mexico working with Professor Arturo Rosenblueth in medical research Professor Wiener moved me from my office Room 2-171 into his office

Room 2-155 to be there while he was in Mexico. It was the beginning of February 1951. I was 20 years old, and the first person ever to occupy the office of Professor Wiener in his absence.

In the office I was alone with Wiener's books and mementos. Although I was in the Mathematics Department, I was in the process of obtaining a Master's Degree in the Economics Department under Professors Paul A. Samuelson and Robert Solow. In the spring of 1951, I was taking the second semester of Samuelson's advanced course in Economic Analysis. At the same time I started writing my SM thesis under Professor Robert Solow. I worked out a mathematical model for the Schumpeter theory of economic innovations. I converted Schumpeter's verbal analysis into equations. I used Wiener's work together with classical time series analysis.

The problem was to find the innovations underlying economic time series. The crucial link was the recognition that the onset of an innovation, such as a new technological advance, cannot be predicted. Thus the onset of each innovation produces a definite and measurable prediction error. I reasoned that the timing of the economic innovations can be found from the economic time series as follows. First compute the prediction operator for the given time series. Next apply the prediction operator to obtain the predicted values. However, the predicted values are not the object of the analysis. Instead, the prediction errors are desired. The prediction errors can be obtained by subtracting the actual values of the time series from the predicted values. These prediction errors represent the desired economic innovations.

Could the same time series method work in exploration geophysics? My plan was to take the digitized seismic traces and treat them as economic time series. Then I would carry out the prediction error filtering. In modern terminology the process is called deconvolution. The prediction error series is the deconvolved time series.

Prof. Wadsworth was on the staff of the Mathematics Department at the Massachusetts Institute of Technology since 1934. His original interest in the field of mathematics was in systems of partial differential equations and the geometry of algebraic pfaffians, but later his interest turned toward the field of mathematical statistics. In the last 10 years he was mainly concerned with problems in Operations Research applied to military and industrial situations and in the application of time series analysis to meteorological data. During World War II, Dr. Wadsworth was a special consultant in the field of weather to the Army Air Forces both in the United States and in the European theatre of operations, and also worked with Ordnance on inspection problems.

Paul Anthony Samuelson is recognized as the father of modern economics. Samuelson was an accomplished mathematician. The first Nobel Prize in economics was awarded in 1969. Paul Samuelsson received the second Noble Prize in economics in 1970 with the citation "for the scientific work through which he has developed static and dynamic economic theory and actively contributed to raising the level of analysis in economic science"

Professor Samuelson and Professor Solow both gave valuable insights to help me in my work. Over the years I always was in touch with Professors Samuelson and Solow. I kept the following letter from Professor Samuelson.

Chapter 1. Norbert Wiener and digital signal processing

MASSACHUSETTS INSTITUTE OF TECHNOLOGY

DEPARTMENT OF ECONOMICS

CAMBRIDGE, MASSACHUSETTS 02139

July 20, 1983

Dr. Enders A. Robinson
100 Autumn Lane
Lincoln, Ma. 01773

Dear Enders:

Congratulations on the recent honor awarded by the Society of Exploration Geophysicists. I have watched the spread of your scientific researches with admiration and I know that Harold Freeman and Bob Solow have also.

Sincerely,

Paul

Paul A. Samuelson

Professor Robert Solow is a professor of economics at MIT. In 1987, he was awarded the Nobel Prize "for his contributions to the theory of economic growth." In 1996 I corresponded with Professor Solow. He wrote back:

> Many thanks for your letter, which brings back memories. Can it really have been almost fifty years ago? The answer is, yes it can. I was in my twenties, you were a mere lad, and even Herman Wold was young and vigorous. Your summary of early deconvolution will be an eye opener to today's students, who have all the computer packages available and take ideas for granted that were new then. I am glad you are making a record of those events, and proud to have played a part in it.

Paul Anthony Samuelson turned 90 in 2005, and MIT and McGraw Hill Co. (his long-time publisher) threw him a party in Boston. He looked as good as ever as did Professor Solow. Back in 1950, when Samuelson entered the classroom, he carefully pulled down all the blinds to shut out the afternoon sun. Economics is called the dismal science, but not in that classroom where Samuelson's light brightly shined.

Origin of deconvolution

When I was a child, my Aunt Ethel would visit our family. She was a piano teacher. A musical score is a written form of a musical composition showing all the notes of a piece of music. Given the score, my aunt could play the music. This is the direct problem (synthesis).

Mozart could listen to a piece of music and then write down the score. This is the inverse problem (analysis). The musical score is a sparse record of discrete symbols, whereas as the music is a copious record of continuous wave motion.

Thomas Edison invented the tin foil cylinder phonograph in 1877. Before Edison, musical scores were recoded but no music or speech was ever recorded. Conceptually the phonograph is arguably a greater invention than the printing press. The press was a technical breakthrough in which symbols could be printed by machine instead of being handwritten. The phonograph was the first instrument that could record the actual sound waves of speech and music.

Let us translate this setting to the seismic case. Each subsurface interface produces a primary reflection. Given a configuration of interfaces, the earth produces a seismic record (which represents wave motion). That is the direct problem. Given a seismic record, the geophysicist by eyeballing finds the primary reflections (which represent discrete interfaces). That is the inverse problem.

I wondered how the inverse seismic problem could be solved. One thing was clear. Because the earth synthesized the seismic record in real time, realizable digital filters should be prominent in the analysis. (Non-realizable digital filters can be realized in real time by introducing a large-enough time delay during which recent data is buffered.) A high-speed digital computer would be mandatory. Neither hand-operated calculators nor analog computers could do the job on a routine basis.

Put yourself in the position of a seismic wave as it make its way down into the depths of the earth. It is like stepping into a completely dark labyrinth. Each time that you bang your head against a wall, you get a sharp jolt. Attached to this sharp jolt are the waves of intense pain reverberating in all directions. The sharp jolt represents the primary reflection. The reverberation of the resulting pain represents the wavelet. The folding together of all this activity is the recorded seismic trace. When you bump your head (the sharp jolt) is unpredictable. The ensuing excruciating pain (the wavelet) is predictable. The computation of the predictive deconvolution filter is based upon this unpredictable-predictable dichotomy.

The word "convolution' means folding. Thus we have the convolutional model: The seismic trace is the convolution of the reflectivity series (the sharp jolts) and the wavelet (the ensuing reverberations). All of this activity occurs in real time. Convolution represents the forward problem.

Deconvolution represents the inverse problem. Deconvolution is the unfolding of the seismic trace in order to detect the jolts (primary reflections) on the seismic traces, just as Mozart could listen to music and then write down the score. The addition benefit of deconvolution is the determination of the seismic wavelet. Deconvolution could be performed by using the equations of Wiener's prediction theory, not as they are used in the prediction of future events, but in an entirely new way.

I worked on the concept of deconvolution for most of the semester. I had the digitized seismic traces in hand. I had the digital signal processing method of deconvolution in principle. The two had to be put together in order to verify that deconvolution worked in practice. Professor George Wadsworth came to the rescue. His assistant Virginia Woodward was an expert on the Marchant hand calculator. Virginia would perform the computations in between her other duties in July and August of 1951. When the numerical results came in, I purposely decided not to get my expectations up. Usually, nothing works on the first try. As I plotted number by number, I was amazed. Deconvolution worked and worked well. It picked out the primary reflection whether they were hidden by reverberations or not. From that moment in August 1951, I knew that deconvolution was the way to detect reflections in a seismic survey.

The mathematical funds available for geophysics in the Mathematics Department were completely exhausted in September 1951. Professor Wadsworth assigned me to meteorological research for which he had funds. It seemed that deconvolution would die in the bud. However, Professor Patrick Hurley of the Department of Geology and Geophysics came to the rescue. Prof Hurley was delighted with the deconvolution results. He was interested. It was like a breath of fresh air. In October 1951, Professor Hurley asked me to write a description of the deconvolution method including a few graphs of the results. He sent the document to Magnolia Petroleum Company in Dallas, Texas.

Early in December 1951 Professor Hurley told me that Magnolia liked the results but they feared that the mathematics could not be could be not implemented for rapid use. As a result they concluded that deconvolution could not be used on a routine basis, and so deconvolution would suffer the fate of other useless mathematical oddities.

I had spent the summers of 1949 and 1950 at Aberdeen Proving Grounds, where I had become familiar with the ENIAC. I instinctively knew that deconvolution could be done on a digital computer. I told Professor Hurley that deconvolution could be used on a routine basis on a digital computer for the processing of seismic records. Within a few days Professor Hurley introduced me to Howard Briscoe, who was a geophysics undergraduate student. Briscoe was friendly with students working at the MIT Digital Computer Laboratory. Briscoe told me that the Laboratory had a digital computer named Whirlwind. Whirlwind and the MIT Digital Computer Laboratory were located in the MIT Barta Building.

Whirlwind had just become operational. In September 1951, the Air Force awarded MIT a contract to carry out Project Lincoln (Lincoln Laboratories) which would conduct research and development in air defense. The centerpiece of this project was the Whirlwind digital computer developed by Jay Forrester in the MIT Servomechanisms Laboratory.

There were not yet any written instructions on the subject of programming for Whirlwind. Briscoe gave me a few pages of a program written for Whirlwind. By studying those sheets I learned how to code. Briscoe and I started writing codes even though we had no way to access Whirlwind in order to test them.

Early in January 1952 Professor Hurley told me the deconvolution project would be funded. Professor Hurley was a hero in my eyes. I never asked but I just assumed that it was Magnolia providing the funding. Professor Hurley told me that the project was to be moved from the Mathematics Department to the Department of Geology and Geophysics. The project would be named The *Geophysical Analysis Group* (GAG). It would begin in February 1952.

The next chapter discloses what went on behind the scenes.

Chapter 2. MIT President James R. Killian

MIT Department of Geology and Geophysics

In 1972 Prof. Stephen Simpson gave me a box that contained old files of the GAG. He said that he had kept these records because he planned to write a history of the GAG. However, his plans had radically changed and he wanted me to have the records so that I could write the history.

September 26, 1951

Dean George R. Harrison
3-207

Dear Dean Harrison:

In the past few months a small amount of exploratory work has been done on the analysis of seismic records, using autocorrelation techniques for the purpose of picking out components that are continuous and predictable and also to explore the possibilities of being able to detect discontinuities due to small reflections from deep structures.

Mr. Enders Robinson, a graduate student in the Mathematics Department, has done much of the work and has analyzed several records supplied by Magnolia Petroleum Corporation. Results of the work have been promising and certainly appear to warrant a continuing project.

It is proposed tentatively that Mr. Dayton H. Clewell of Magnolia Corporation be approached with a request for $15,000 as a grant-in-aid to support Mr. Robinson and minor computing assistance for a period of time necessary to investigate more thoroughly the meaning of the present findings. The request would include an understanding that Magnolia would co-operate in supplying records of a special type.

It is proposed that the project be under the geophysics section of the Department of Geology, under Mr. Robinson's supervision with Doctors Wadsworth, Haskell, and myself as an active working committee to direct the research. We would be very glad if you would give us your opinion on this.

Respectfully yours,

P. M. Hurley

cc: R. R. Shrock
G. P. Wadsworth
N. H. Haskell
E. Robinson

I received a copy of the above letter. I did not receive a copy any of the following letters at the time, because as a student I was not privy to financial information. However, I was apprised of any of the scientific content that was pertinent to my work.

OFFICE OF THE DEAN OF SCIENCE

October 1, 1951

Professor P. M. Hurley
Room 24-416

Dear Professor Hurley:

The project outlined in your letter of September 26 seems to me to be one well worth prosecuting, and I favor making application to the Magnolia Corporation for a $15,000 grant-in-aid as you suggest. I am referring this matter to Professor Walter Gale, Secretary of the Institute, for coordinating with the Development and Liaison offices, and he will doubtless be getting in touch with Professor Shrock and yourself in connection with the preparation of a suitable application. The matter of Institute overhead should be considered very carefully in connection with an application of this sort, and matters of space, personnel, and so on should be laid out very carefully before the application is forwarded from Professor Gale's office, to make sure that we can fulfill any commitments that would be implicit in the application.

Yours sincerely,

George R. Harrison

GRH:LD

cc: Professor W. H. Gale
    Professor R. R. Shrock

On November 1, 1951 Professor Hurley sent a letter to Magnolia. Appended to the letter was a summary statement which I wrote up at his request. His letter and my statement follow.

November 1, 1951

Dr. David H Clewell
Magnolia Petroleum Company
PO Box 900
Dallas, Texas

Dear Dr. Clewell:

Dr. G. P. Wadsworth of the Mathematics Department at MIT has continued with his active interest in investigation of the characteristics of the noise in exploration seismographs by use of auto- and cross-correlation techniques. During the summer, he has had two people [Added note: The second person was Virginia Woodward, who did the computations on a Marchant calculator] working on this, supported by funds from the Institute. Also the work is been discussed in several conferences which Dr. Wadsworth invited doctors N. H. Haskell, J. G. Brian and myself, to consider the meaning of the results. It appears that there is considerable amount of predictability in the responses shown on the seismograms that you sent several months ago, and the reflections stand out rather sharply as unpredictable components in the record. A summary statement of the working findings today is appended.

As a result of the preliminary findings, it has been decided Mr. Enders A. Robinson, the mathematician who carried out most of the work this summer, be engaged as a Research Associate in the Geology Department to continue the investigation, and that Drs. Wadsworth, Haskell and myself act as a working committee to supervise the program. If obtaining support for the work, it was decided that we should approach you and your company for continued aid in supplying data in the form of carefully controlled seismic records, and also for financial support to cover the basic cost of this investigation for a period of time sufficient to convince ourselves that there is really something of interest in this program. We have taken this matter up with the Dean of Science and the Administration, and have received approval in seeking a grant-in-aid of $15,000. This amount is the cover the salary of Mr. Robinson for about two years and minor computing assistance, with an amount left over that we estimate not to exceed $2,000 for the purchase of a piece of equipment that will be useful to solution some of the simultaneous equations.

Mr. Robinson will be free to start work full-time after the first of February 1952, and is quite important that we engage his services by this date if we wish to keep them here. As a result, any decisions that are made in this project must be fairly immediate. I had hoped to have a chance to come to Dallas to see the people at GSI before now, and at the same time approach you informally regarding this proposal. However, I am afraid that the trip will continue to be delayed, so we are enclosing a brief statement regarding the present status of the work in the hope that it may be sufficient for a decision in your part, since you are already well acquainted with the ideas involved.

Sincerely yours,

P. M. Hurley

Appended is a statement by Enders A. Robinson:

Summary of the work done on the Analysis of Seismic Records utilizing the techniques of generalized harmonic analysis

The purpose of the study carried on at MIT during the summer of 1951 was to answer a few specific questions regarding the behavior of seismic records considered as time series. The data used were eight records each resulting from a single shot and consisting of four traces produced by the output of 4 geophones placed along the line with the center at the shot-point. These data were supplied by Magnolia Petroleum Corp. The questions to be answered were:

1. Are the individual records predictable sometime in advance of a fixed point t_0 assuming that only the data before t_0 is known.

2. Does the linear operator which constitutes the prediction mechanism change as a function of time over the length of the record?

3. How does the spectrum of the frequencies and their phase relationships vary as a function of time from the time that the explosion occurred?

4. Is it possible to determine the points where a deep reflection occurs in the records through purely statistical means by observing a change in the dynamics of the system at that point?

The tentative answers to the four questions based entirely upon the analysis of the aforementioned records are as follows:

1. It was found that the records were very predictable in a statistical sense up to several units in advance utilizing a very simple linear operator involvement past of the traces to be predicted and the past of one or more of the simultaneous traces taken in the same experiment. Because of the amount of work involved, it was impossible to determine how much of the past an operator should incorporate in order to predict quite a distance into the future, but it might be stated with the series themselves seemed to possess the qualities of very predictable time series. The fact that the stable

dynamics occurred for a reasonably long period of time seems to rule out the possibility of phenomena being random.

2. The linear operator does change gradually over the length of the record, particularly during the period immediately following the explosion. It does, however, remain reasonably constant over a long enough span of data to study it effectively after the initial disturbance is over with.

3. Careful attention was given to the behavior of the spectrum and the phase relationships between the various frequencies as time progressed from the shot-point to the end of the records. They seem to be enough change in both the fundamental spectra of each of the series and the change in the phase relationships indicated that the cross-correlations to assume that the predictability was not produced by the effect of random impulses on a sluggish oscillator, thus eliminating the possibility that this predictability is due to the geophone response. It may be possible eventually to tie these shifts to the structure the ground at lower levels.

4. Linear operators were constructed for the data just prior to the points of the deep reflections as indicated on the records. The operators were then utilized to predict the period during which the reflection occurred. In every case the errors of prediction increased very markedly during this period and after the interval of time during which the reflection occurred the errors dropped back to approximately their value which was previously demonstrated for the period during which the operator was developed. On several of the records the deep reflection failed to occur as it had in most of the seismograms, but nevertheless, in these instances the errors of prediction jumped up in the same manner. Although this is not conclusive, it seems to indicate that during the time of reflection an entirely different set of dynamics is introduced into the system. If this is true, and introduces a new approach into the problem of determining when additional energy is introduced into the system through reflections from a deep layer.

Enclosed is a graph indicating the average errors of prediction previous to the reflection, during the reflection immediate after the reflection and beginning of a new reflection.

The answering letter is

Chapter 2. MIT President James R. Killian

MAGNOLIA PETROLEUM COMPANY
A SOCONY-VACUUM COMPANY
FIELD RESEARCH LABORATORIES
P. O. BOX 900
DALLAS 1, TEXAS

T. W. NELSON
DIRECTOR

D. H. CLEWELL
ASSISTANT DIRECTOR

DUNCANVILLE ROAD
BETWEEN LEDBETTER
DRIVE AND FIVE MILE ROAD

November 29, 1951

Professor P. M. Hurley
Department of Geology
Massachusetts Institute of Technology
Cambridge 39, Massachusetts

Dear Professor Hurley:

      Your letter of November 1 has been circulated here at the Laboratories, and the results you presented are distinctly of interest. They represent substantial progress since Ed White's visit with Professor Wadsworth last year. The curve you sent shows excellent "signal-to-noise" for a single trace, and a multiplicity of curves derived from geophone signals in a conventional reflection spread would appear to show great promise. We have no very good appreciation of the mechanical operations required for predicting the geophone signals and displaying the difference between predicted and measured signals. Unless instrumentation can be developed to do this rapidly, the method would probably be difficult to apply to commercial prospecting. Is it correct to assume that the operations stand a chance of being instrumented for rapid use?

      With respect to your request for a grant of $15,000 grant-in-aid (over a two year period) to pursue this work further, we wish to call to your attention that we, as an affiliate of the Socony-Vacuum Oil Company, Inc., are already contributing to M.I.T. to the extent of $50,000 per year. We believe the work that you propose to do is very worthwhile and a proper next step. Accordingly, it is our thought that the funds you require might properly be taken from the contribution we are now making since your work is of immediate and direct interest to us. We are not familiar with the detailed conditions under which our present contribution of $50,000 a year is made, but perhaps you could inquire about the possibility of securing funds in this manner from the proper persons at M.I.T. In the meantime we are taking the matter up with our New York people to ascertain the possibility of having Socony request that your project be supported from our present contribution.

      In the event that you might wish to correspond directly with our New York people on this matter you should address such correspondence to Mr. W. M. Holaday, Socony-Vacuum Oil Company, Inc., 26 Broadway, New York, N. Y.

      I missed you last summer when the M.I.T. students were here and regret that your trip is still delayed. Please do visit the Laboratories whenever you can.

Very truly yours,

D. H. Clewell

DHC/lc

cc - Messrs. W. M. Holaday
              Henry C. Cortes

Prof. Hurley conveyed to me the scientific content of the above letter. Two things stood out in my mind. The first was "the curve shows excellent signal to noise ratio." The second was "unless instrumentation can be developed to do this readily, the method would be difficult to apply to commercial prospecting."

As related in Chapter 1, I told Professor Hurley that a digital computer could do the processing of seismic records. As a result Professor Hurley introduced me to Howard Briscoe, who knew about the MIT Digital Computer named Whirlwind which had just become operational.

Thomas Alva Edison invented the first long-lasting, practical electric light bulb. As such it was useless to most people as they had no electric power source. As a result, Edison developed a system of electric-power generation and distribution to homes, businesses, and factories, which was a vital development in the industrialized world. His first power station was at 255 Pearl Street in Manhattan, New York.

In effect, Clewell was saying that deconvolution looks good, but it is useless unless you find a way to do it commercially. Clewell was right. We had to build, in effect, a "power station," as it were, so that digital signal processing could be used in commercial prospecting. The "power station" would have to be the digital computer.

The next letter is:

> December 6, 1951
> Prof. Robert R. Shrock
> 24-302
>
> Dear Professor Shrock:
>
> We have had a reply from Dr. Dayton Clewell of the Magnolia Corporation regarding our request for $15,000 to initiate the seismic record program. Copies of the original letter and reply are attached. In his reply, Dr. Clewell stated that his company was interested in the work and suggested that funds he taken from the annual $50,000 given by the parent company (Socony-Vacuum) to the Development Program.
>
> As you remember in our meeting with Professor Gale, when considering the effect of approaching Magnolia at the outset, it was decided that overhead should not be included in the budget in view

Chapter 2. MIT President James R. Killian

of this Company's present support. In effect, therefore, the cost to the Institute for this work would be substantially in excess of $15,000. In your absence and at your request yesterday, I took this matter to Prof. Gale and I believe it is now under consideration by his office.

In attempting to look ahead at the possible future of this method of attack, Prof. Wadsworth and I believe that it is one of these things that may either die after about one year of further research, or else be useable in seismic exploration work. In the latter event it will be immediately and definitely of value to petroleum companies. We feel that this is a gamble that should be supported by the oil companies, and we are disappointed that if the work is to be furthered, it appears necessary that the Institute draw upon its development funds to support it.

No budget was submitted in the original proposal, but if the job were to be done in the most efficient way, the estimated $15,000 might be cut to $12,000 in a period of 1 1/2 years, without overhead allowance.

Yours respectfully,

P. M. Hurley

The fifty thousand dollars a year that Socony-Vacuum was giving MIT went mostly to the Chemical Engineering Department for research in refining methods. This research led to the production of a greater amount of valuable components such as gasoline from each barrel of crude oil. These were long term research projects that could not be disrupted. There was an impasse that could not be overcome.

The Chemical Engineering Department did not want to part with money already given to them. Socony-Vacuum did not want to give any new money for the deconvolution project. The deconvolution project was dead in the water. However, Prof. William Ted Martin (head of the Mathematics Department), Prof. George Wadsworth, Prof. Norbert Wiener and Prof. Norman Levinson were solidly behind the project.

Christmas present in 1951

James Rhyne Killian, Jr. (1904–1988) was the president of MIT. At this critical juncture, President Killian made the vital decision. On December

26, 1951, President Killian sent the following letter to Socony-Vacuum. (I did not know about this letter at the time.)

> December 26, 1951
>
> Mr. W. M. Holaday
> Director of Research
> Socony-Vacuum Laboratories
> 26 Broadway
> New York 4, New York
>
> Dear Mr. Holaday:
>
> This is with reference to the subject of Dr. Clewell's letter to our Professor Hurley dated November 29, and of Dr. Killingsworth's telephone conversation with our Dr. Bevans of December 6.
>
> In accordance with the latter discussions **we are proceeding on the basis of allocating some of our general purpose funds to the development of this program on the analysis of geophone responses.** You are aware of the general financial outlook for institutions such as M.I.T., and for that reason I am sure will appreciate our wish not to deviate from the general principle of reserving income through the Industrial Liaison Program for support of our existing activities, and not employing these funds for any projected expansions. We feel that if we begin to accede to requests for diversion of these funds from existing activities to new ones, such practices will tend to grow and perhaps be irreversible. This would of course defeat our objectives in acquiring such general support.
>
> This additional activity will not at its inception be a large one, and for this reason if industrial support is not readily forthcoming we are sometimes thus able to arrange for an internal appropriation. However, if after 18 or 24 months, the activity shows even more promise and thus warrants further expansion, I know outside support will be a definite requirement.
>
> In view of the close cooperation that is already taking place between your Magnolia staff and our own faculty in this area, if the program develops favorably in this way we should like to provide Socony-Vacuum first with the opportunity to render such additional support at that time. In the meantime, needless to say, your staff will be kept fully informed on the progress we are making in this field.
>
> We hope this course of action meets with your full approval.
>
> Yours sincerely,

J. R. Killian, Jr., President
cc: Dr. Clewell
Dr. Killingsworth
Prof. Hurley

This letter of President J. R. Killian was decisive in establishing the presence of digital signal processing and deconvolution. The MIT President's Office funded the deconvolution project for 1952 and some of 1953.

In January 1952, Professor Hurley introduced me to Ted Madden who was the only geophysicist on the MIT staff and to Dr. Normal Haskell, who was on the staff of the Air Force Cambridge Research Center but taught geophysics courses at MIT. Professor Hurley scheduled an air trip to Dallas, Texas in early February 1952 so I could meet the geophysicists at Magnolia and also at Geophysical Services Inc. (the exploration company of Texas Instruments). However my father died of cancer on February 4, 1952, so the trip was cancelled.

In order to obtain outside support, it was deemed essential to provide an operational answer to Clewell's question of November 29, 1951. The task was clear. Signal processing had to be implemented on a digital computer. The Air Force had set aside some time on Whirlwind for academic use. President Killian gave us permission to use Whirlwind and we became the first major academic user. I was determined to demonstrate that deconvolution could be performed on a production basis.

History of Whirlwind

By 1947, Jay Forrester and his associate Robert Everett at MIT started work on the design of a digital computer which they called Whirlwind. In 2011, MIT celebrated its 150th anniversary. On October 15, 2011 the MIT had an event entitled "Project Whirlwind." The primary focus was a session with Jay Forrester and Robert Everett. A couple dozen participants from the Whirlwind development era in the late 1940s and 1950s were present for the presentation.

Everett commented that their original intention was to build a machine with a 32-bit word length, but instead they built half a machine— one with a 16-bit word length. Most computers of the era operated in bit-serial mode, using single-bit arithmetic and feeding in large words, often 48 or 60 bits in size, one bit at a time. This was simply not fast enough for their purposes, so Whirlwind with its bit-parallel mode operated on a complete 16-bit word in every cycle. Ignoring memory speed, Whirlwind in 1951 with 20,000 single-address operations per second was the fastest digital computer of its time.

In the questioning phase of the session, Forrester was asked about software and the overall system rational, as opposed to the hardware that he and Everett had mostly been describing. Forrester explained that they had to develop a large team of programmers that had not existed before. From the audience, John Frankovich (an early software innovator on Whirlwind) said that in the years after Whirlwind became fully operational, students and others from around the campus developed lots of software, including the first algebraic compiler. (John Frankovitch was a close friend. We were undergraduate students together in the Mathematics Department in the class of 1950.)

Forrester then noted that a number of new fields got started based on Whirlwind such as **oil field exploration analysis**. Frankovich added that pioneering work was done on Whirlwind in the fields of numerical milling machine control, studies of radio station radiation patterns, TV frequency assignment calculations, photographic lens design, and so forth. It is interesting to note that in 2011 Forrester chose oil field exploration analysis as his prime example of a field that got started based on Whirlwind.

Deconvolution implemented on Whirlwind 1952

In February 1952 I transferred to the Department of Geology and Geophysics with the appointment as Research Associate. The Geophysical Analysis Group (GAG) officially came into existence at that time. Stephen Simpson was a graduate student in geophysics who had received his undergraduate degree in physics at Yale. Simpson joined the GAG as a research assistant. Howard Briscoe was an undergraduate, and worked part time on the GAG.

One of my first tasks was to teach Stephen Simpson how to write codes for the Whirlwind computer. He was a little reluctant at first, but all of a sudden the computer bug grasped him. He became one of the foremost programmers at MIT, and a strong advocate of computers and artificial intelligence. He was already involved with his own material for his PhD thesis, and he immediately launched into using Whirlwind to do the computations involved. He was always supportive and took an active interest in managing the GAG. He was a good and life-long friend.

Simpson always followed his own directions and did many different computer-related things. He never worked on deconvolution. In contrast, I spent all of my efforts on a narrow path. My purpose was to deconvolve enough seismic signals so we could be confident that deconvolution was efficacious and robust. Only then could deconvolution be used in the discovery of oil and natural gas. No research assistant on the GAG was ever told to follow that narrow path. Some did and others went their own way. However, it was mandatory for all research assistants to learn how to use computers.

In the spring of 1952, Briscoe and I continued working on the Whirlwind codes for deconvolution. We had the five months from February to July 1952 to deconvolve more seismic records on Whirlwind. The oil companies were using analog filters. We could imitate their analog filters. We Fourier transformed the desired frequency characteristics to compute the numerical filter coefficients. The filtering was done in the time domain by a convolution code. Whirlwind was ideally suited to convolution. However, the use of time domain filters *in itself* would not impress the oil companies at all.

It was difficult, sometimes impossible, to convince anyone that a list of numbers (the filter coefficients) could do what an analog electric frequency filter would do. And if someone were convinced, then the reply would be that it was a matter of cost. Whirlwind occupied a whole building and required a large staff to maintain it. Why would an oil company want to go to all of the trouble of buying an expensive digital computer in order to duplicate what they were already doing quite well with inexpensive analog filters?

With deconvolution we could go beyond what the oil companies could do. Deconvolution was a process of geophysical inversion. It required that the filters be designed on the basis of the data at hand. Analog devices could not do inversion, but a digital computer could.

The deconvolution problem consisted of two parts. Part 1 consisted of computing correlation coefficients and the solving a set of simultaneous equations in order to obtain the coefficients of the deconvolution filter. Part 1 ideally required a large word size, something other computers had but not Whirlwind.

Part 2 consisted of using the deconvolution filter on the seismic trace to obtain the deconvolved seismic trace. Whirlwind with its high speed and small word size was perfect for do Part 2. I soon had a digital filtering program working on Whirlwind that would do Part 2; namely, deconvolution of the seismic traces. However, before this program could be used we had to write a program that would do Part 1; namely, computation of the numerical coefficients of the deconvolution filter. Therein was the difficulty.

In the spring of 1952 Whirlwind was in its original embryonic state. It was in the process of being tested and accordingly improved. Hardware was the objective, not software. It was a vacuum tube machine; the transistor was yet to be perfected for computers. The internal memory was electrostatic storage, fairly unreliable but the best at the time. Whirlwind in the spring of 1952 had random-access electrostatic storage of 1024 (i.e., 1K) words of 2 bytes each. In other words, the RAM was 2 kilobytes. Today the single word "Whirlwind" stored in Microsoft Word requires 16.4 kilobytes. About a year and a half later, the RAM on Whirlwind was doubled to 4 kilobytes. It was paradise.

When I first entered the Digital Computer Laboratory in February 1952, I asked for the programs for solving simultaneous equations. When they said there were none, I got a sinking feeling. The hard fact was that there were no codes yet written that would do the kind of mathematics that we needed. The programmers working at the MIT Digital Computer Laboratory were writing codes not for mathematics, but for housekeeping operations on the computer. We had to write all the codes for digital signal processing from scratch.

In the construction of Whirlwind little attention had been given to software compared to that given to the designing the computer hardware. The design engineers believed that the task of writing the required software could be easily done upon completion of the hardware. The fact is that Whirlwind was not a computer built to solve mathematical problems. It was built to transmit data quickly from dispersed radar sites to a central station. Essentially no numerical mathematics was required for data transfer. In contrast digital signal processing requires the use of heavy numerical mathematics.

Whirlwind was a general purpose computer, but its intrinsic structure (such as word size) was not chosen to do extensive numerical calculations. It had no floating point arithmetic. In fact, it was a highly specialized computer for the rapid transfer of low precision numbers from various radar sites for display on a monitor at a central station. By July 1952 Briscoe and I did succeed in writing a double precision program that would fit in Whirlwind's limited storage space. The word "double" derives from the fact that a double-precision number uses twice as many bits as a regular number. On Whirlwind a single-precision number required 16 bits. Its double-precision counterpart was 32 bits. In other words, a double-precision number required two 16-bit words. Our program by necessity had to be rudimentary. Because of its limited memory, it could not have the all of the accuracy that we would desire. The Levinson recursion was designed for the case of a single input signal. Because we were using two or more input signals, we inverted the matrices by conventional means. Some years later we did expand the Levinson recursion so it would handle the case of several input signals.

In doing Part 1, everything was done in carefully scaled integers, avoiding any unnecessary division. If our program failed to keep the numbers in a specific problem small enough, there would be overflow, and the program would crash. In order to keep the numbers small, we would scale and round-off the numbers. Too much round-off would make the autocorrelation matrix singular. In such a case we could not invert the matrix. However, we could usually make the matrix invertible by adding a small positive number to its diagonal. This method is even in use today under the designation of pre-whitening. In recalcitrant cases, we would carefully check our results with the Marchant desk calculator.

We demonstrated that we could do the digital signal processing of seismic records on the MIT Whirlwind digital computer. Recall that in his letter dated November 29, 1951, Clewell wrote, "Unless instrumentation can be developed to do this rapidly, the method would probably be difficult to apply to commercial prospecting." We now had shown that the instrumentation for deconvolution was the digital computer. The next step was to present our findings to the oil industry.

The MIT meeting of August 1952

INDUSTRIAL LIAISON OFFICE

July 25, 1952

Memorandum to: G. P. Wadsworth
                P. M. Hurley
                E. A. Robinson

We received an excellent response to our letters of invitation to the August 6 conference. There has not been a single turn-down by an oil company and also at least two electronic concerns are to be represented. Among those expected are Dr. Rust of Humble, Dr. Yost of Magnolia, Dr. Herzog of Texaco and Dr. Silverman of Stanolind. Incidentally, each of these men is being accompanied by another from the company.

We are planning on a luncheon at the Faculty Club on that day.

M. V. Barts.

The MIT Industrial Liaison Office arranged a meeting to be held on August 6, 1952. The meeting was entitled "A Conference on the Generalized Harmonic Analysis of Seismograms." The results of everything done to date were presented. Two reports were given to each one present. Those attending were representatives from Magnolia Petroleum and 11 other oil companies as well as from Texas Instruments and United Geophysical.

INDUSTRIAL LIAISON OFFICE

August 8, 1952

Memorandum to: Prof. G. P. Wadsworth
Prof. P. M. Hurley
Mr. E. A. Robinson

This is just a note of appreciation from our office for all your efforts in connection with Wednesday's program. We think it quite obvious that the majority attending were impressed with the concepts and material presented. The occasion proved to be, as we were sure it would, a significant one in the Industrial Liaison Program.

We hope this office can, in turn, be of assistance to you in making arrangements for the cooperative program spontaneously suggested by the group which met again Thursday morning.

With best regards.

H. V. Bartz

44 MIT AND THE BIRTH OF DIGITAL SIGNAL PROCESSING

The MIT meeting of August 6, 1952 presented things that the geophysicist had never seen before. They saw so-called "no reflection" seismograms yield the primary reflections. There was unanimous sentiment that their companies should sponsor this research at the MIT Department of Geology and Geophysics.

One direct result of this MIT meeting was the invention of vibroseis.

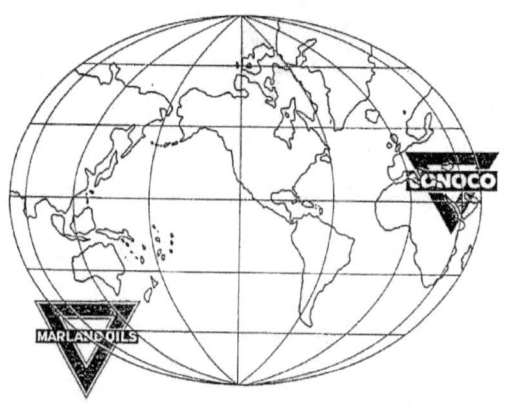

GEOPHYSICS
THE FIRST FIFTY YEARS

John Crawford   Bill Doty

Co-inventors of "VIBROSEIS"

On August 6, 1952, W. E. N. (Bill) Doty attended the symposium, "Harmonic Analysis of Seismograms," presented at the Massachusetts Institute of Technology in Boston. Bill came away from this meeting with a firm feeling that the Information Theory Technique which had been described could be beneficially applied to Seismic Exploration. Returning to Ponca City, he immediately expressed this opinion to John Crawford. John concurred with Bill and the two of them set out to determine a practical means whereby cross-correlation could be applied to seismic signals. The first breakthrough came with the suggestion by John on August 18 that a "sweep signal that is, one which continually changes frequency in one direction during transmission, would provide the desired non-repetitive, long-duration sign Thus, cross correlation of the transmitted and received signals could prov the basic requirement of the seismic system; i.e., travel time between tra mission and reception of a signal.

In the fall of 1952, it was clear that we would need a large amount of

Chapter 2. MIT President James R. Killian

machine time on Whirlwind in order to do digital signal processing. I applied for five hours per week; only one hour a week was granted. I looked for an alternative, and found Raytheon. The Raytheon Manufacturing Company in Waltham, Massachusetts had recently established its Computer Services Section. Richard F. Clippinger was the Head of the Section. With him were Bernard Dimsdale and Joseph H. Levin. Before entering Raytheon, Clippinger had been in charge of the ENIAC, EDVAC, and ORDVAC computers at Aberdeen. Dimsdale had been responsible for numerical analysis at Aberdeen. Levin, formerly at the ENIAC, had been in charge of the SEAC computer at the National Bureau of Standards.

By February 1953 all of the oil and geophysical companies present at the MIT conference of August 6, 1852 had joined in the venture. The project was called the Geophysical Analysis Group (GAG). One or two representatives from each sponsoring company made up the Advisory Committee. Two meetings of the GAG with the Advisory Committee were to take place each year; the winter meeting would be in the Southwest and the summer meeting at MIT.

The MIT GAG was a consortium in which a university (in this case, MIT) would carry out specific research for a number of industrial companies. It was a model for other such consortiums formed in succeeding years. Today much of the research in geophysical exploration is done by such university-industrial consortiums. This system gives graduate students a means of support. It also gives small companies an advantage, because often a small company could not afford to do a large research project on its own. It also gives large companies an advantage, because they can benefit from the independent ideas provided by many small companies.

Raytheon in 1952

In the fall of 1952, hardware improvements were still being made in Whirlwind so the machine had a lot of down days. Radom Access Memory (RAM) was the weak point of every digital computer at that time, except the highly reliable ENIAC. The ENIAC had Read-Only-Memory (ROM) and a bank of accumulators that acted as RAM.

The maintenance time on just Whirlwind's electrostatic RAM was about four hours per day, and the mean time between memory failures was two hours. This situation was rectified a couple of years later when Forrester's invention of magnetic core memory was installed on Whirlwind. This development made Whirlwind the first reliable RAM computer.

Forrester had lined up various companies to participate in design studies for prototypes of the production model of Whirlwind. One of the companies would then be chosen to manufacture the production model. The three companies considered as the possible manufacturer were IBM, Raytheon, and Remington Rand. IBM was chosen. This gave IBM the opportunity to use the Whirlwind innovations, especially its magnetic core memory, in their own computers. It was to the advantage of IBM, for they soon became the leading computer manufacturer.

At the August 6, 1952 conference, we demonstrated by numerical examples that Whirlwind did every step in the deconvolution of digital seismic records. However, we pointed out that the 16-bit word size of Whirlwind made it unattractive for Part 1 of the computations (i.e., the least squares determination of the deconvolution filter coefficients). There was little use in trying to squeeze more out of Whirlwind in that regard, because every other computer had much larger word size. However, we also pointed out that the 16-bit word size of Whirlwind made it attractive for Part 2 of the computations (i.e., digital filtering with the coefficients obtained in Part 1)

We faced another task. Whirlwind was a military computer. Fortunately some time on Whirlwind was made available to the MIT faculty. We were using up a lot of that time, but this could not last. More and more of the MIT faculty were beginning to use Whirlwind. More importantly, no time on Whirlwind could ever be used by outsiders such as the oil companies. I made the decision to bring in Raytheon to provide the computer and the coding to do digital signal processing. In this way an oil company on its own could employ Raytheon to do digital signal processing on proprietary data.

The following is an advertisement by Raytheon in September 1952.

In exploring the possibilities of automatic control as applied to any process, operation or activity, may we remind you that Raytheon has an established position in this rapidly expanding field. Raytheon is now producing digital and analogical computers and other data processing systems, tape handling mechanisms, transducers, microwave links, telemetering systems, programming devices, and servomechanisms.

The Raytheon Manufacturing Company in Waltham, Massachusetts had recently established its Computer Services Section. The three top mathematicians at the ENIAC had left Aberdeen and joined Raytheon to set up the Raytheon Computer Laboratory in Massachusetts. The GAG became the first major customer of the Raytheon Computer Laboratory.

The three Raytheon mathematicians were Dr. Richard F. Clippinger, Dr. Bernard Dimsdale, and Dr. Joseph H. Levin. They had more working experience in digital computing than anyone else. For years, they had worked closely with John von Neumann on a number of mathematical problems on the ENIAC. However, they had only worked on conventional problems in numerical analysis. They were unaware of digital signal processing.

Clippinger, Dimsdale and Levin represented the mathematics of John von Neumann. We gave them our Whirlwind codes for guidance. Our codes represented the mathematics of Norbert Wiener. The resulting set of codes represented the digital union of Norbert Wiener and John von Neumann. Ironically this set of codes was designed for, and used on, the Ferranti digital computer, whose coding system was designed by

Alan Turing. MIT and Raytheon merged Wiener, von Neumann and Turing to form the first commercially available digital signal processing.

The GAG had its first meeting with Raytheon on October 17, 1952. On October 28, 1952 Raytheon submitted an estimate.

RAYTHEON MANUFACTURING COMPANY
EQUIPMENT ENGINEERING DIVISION
148 CALIFORNIA STREET • NEWTON 58, MASS.

October 28, 1952

Dr. Enders A. Robinson, Director
Geophysics Analysis Group
Room 24-033
Massachusetts Institute of Technology
Cambridge 39, Massachusetts

Dear Dr. Robinson:

    In accordance with our agreement with you, we are pleased to submit herewith the following quotation for services which could be performed by our Computing Services Section involving seismogram analysis.

    The estimates are made on the basis of the following statement of the problem:

Phase I
Given several seismogram traces whose ordinates are x_i, y_i, z_i, etc., it is desired to fit the linear operator

$$c + \sum_{s=1}^{4} a_s x_{j-s} + \sum_{1}^{4} b_s y_{j-s} + \sum_{1}^{4} c_s z_{j-s} + \ldots = \hat{x}_{j+k},$$

over an interval containing 54 values of i in such a way as to minimize $\sum_{j=1}^{50} (x_j - \hat{x}_j)^2$. This is to be done for one or three values of k.

Phase II
For each of 39 values of m it is desired to compute and print

$$E_m = \sum_{j=1}^{20} (x_{10m+i} - \hat{x}_{10m+i})^2.$$

Chapter 2. MIT President James R. Killian

Dr. E. A. Robinson   Page 2   October 28, 1952

      The normal matrix to which the above problem gives rise has dimensions $(4t \neq 1) \times (4t \neq 1)$, where t is the number of traces used. Each element of the matrix is an inner product of two fifty-element vectors.

      The quotations are based further on the assumptions that input data are given to two significant figures, output information has three significant figures, the computations include checks of the solutions obtained for the normal equations, data are transmitted to us on a monthly basis with no fewer than sixty (60) cases to be submitted at one time nor fewer than one thousand (1000) cases to be submitted in total. A "case" is defined to consist of the computations for a single set of traces, and the specified number (one or three) of values of k.

      A "Half Job" is defined to consist of Phase I only, and a "Full Job" is defined to consist of both Phases I and II.

Number of Cases	One Value of k				Three Values of k			
	Half Job		Full Job		Half Job		Full Job	
	$t=2$	$t=4$	$t=2$	$t=4$	$t=2$	$t=4$	$t=2$	$t=4$
1000	$5,010	7,740	9,660	14,150	5,380	8,150	12,400	19,500
2000	7,020	10,080	13,920	22,900	7,760	10,900	19,400	33,600
5000	9,450	17,100	26,700	49,150	11,300	19,150	40,400	75,900

      The above may be considered as a firm price quotation for a period of sixty (60) days from the above date, terms net thirty (30) days.

      We trust the above will warrant your further consideration, and if we can be of further service, please feel free to call upon us at your convenience.

      Very truly yours,

      RAYTHEON MANUFACTURING COMPANY

      T. R. Porter, Manager
      Technical Sales

TRP:DG

On November 25, 1952, MIT appropriated $1,000 for Raytheon to begin the necessary coding study. The purchase order read: "Computing services covering coding study and preliminary code for seismogram analysis program. Deliver flow diagram and preliminary code on or before Jan. 1, 1953. Amount not to exceed $1,000.00. Purchased from

Raytheon Manufacturing Company. T. R. Porter, Manager, Technical Sales." On December 23, 1952 Raytheon delivered the flow diagrams with code for digital signal processing and deconvolution; they are in the handwriting of R. F. Clippinger and are dated Dec. 17, 18 and 19, 1952, and initialed (upper right) "RFC." Three pages are given below.

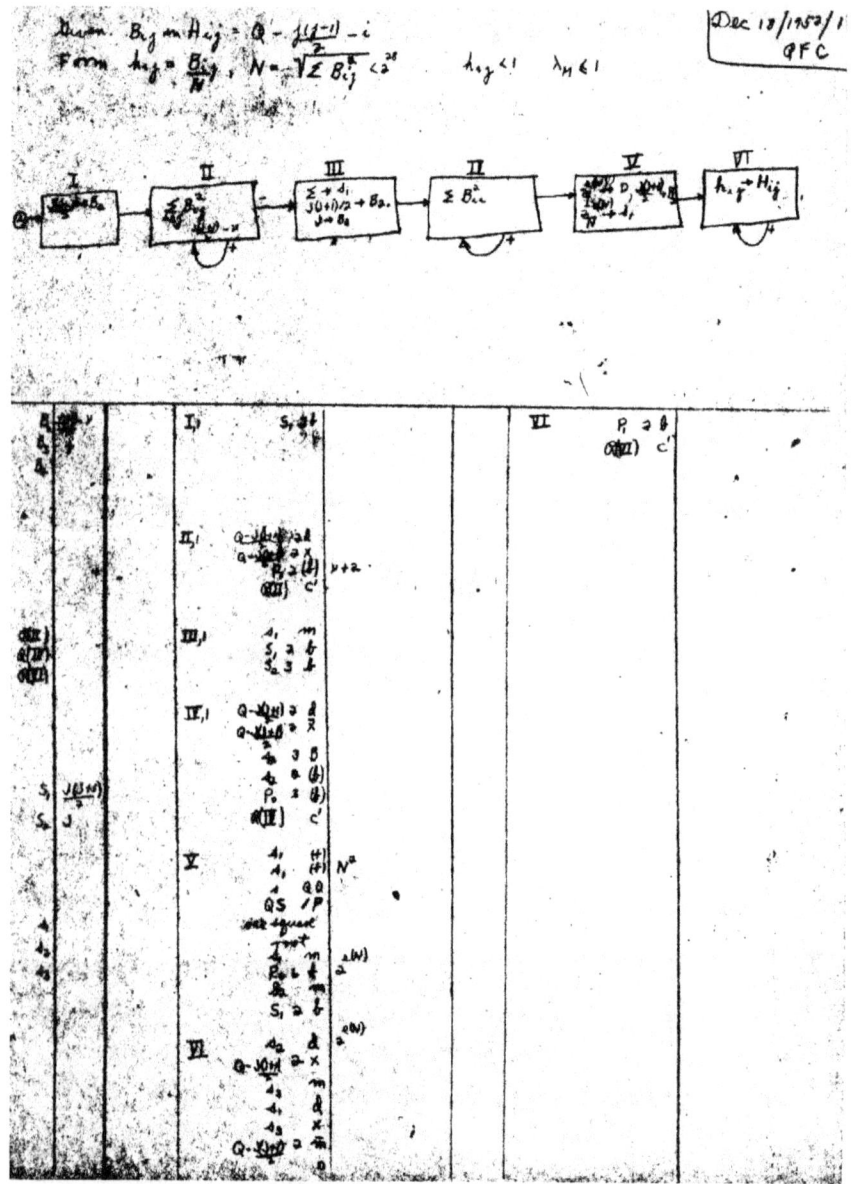

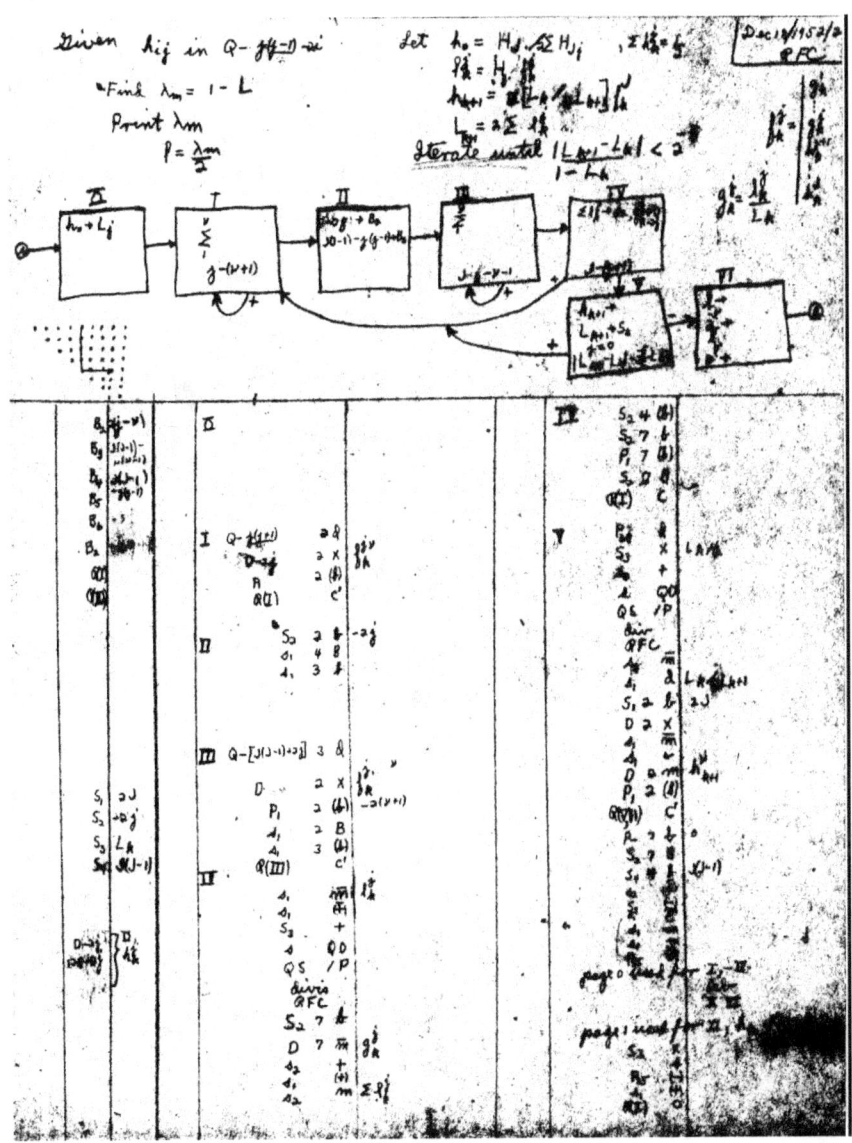

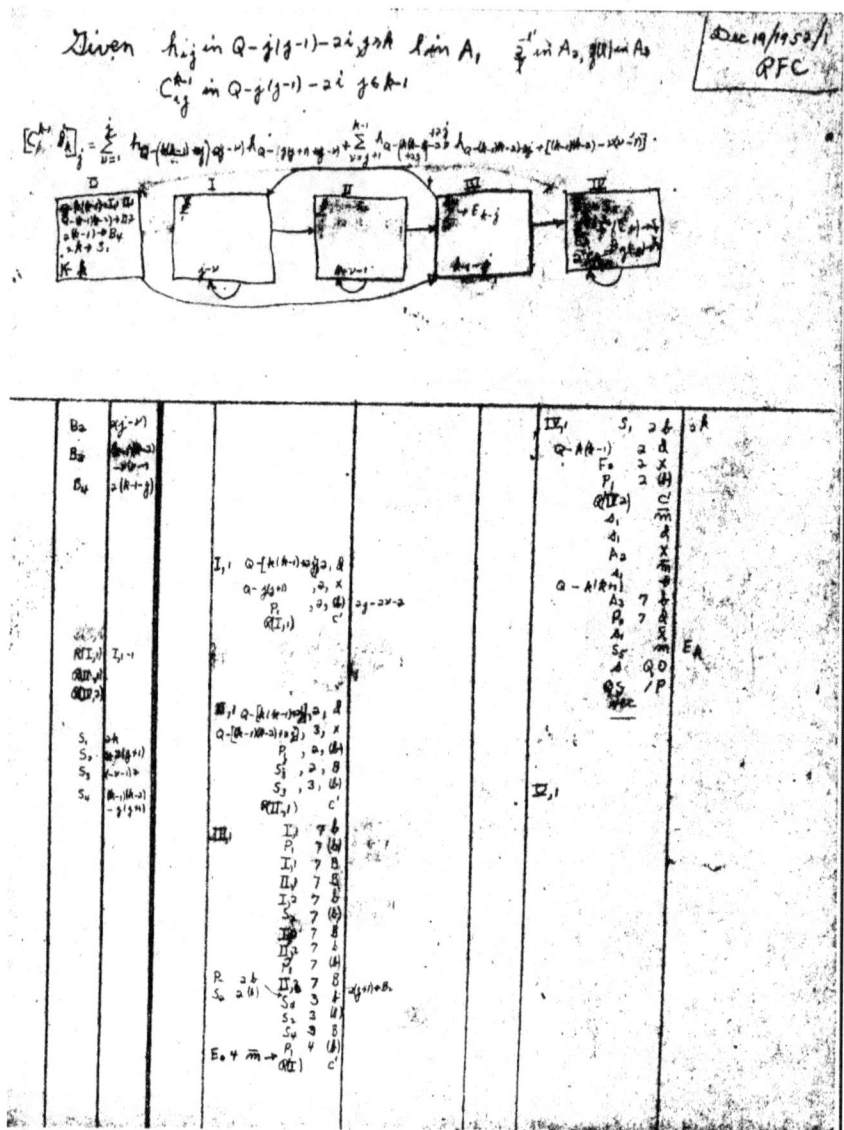

In the 1960s Raytheon bought the oil exploration company Seismograph Service Corporation. It became a leader in digital signal processing. In 1969 *The Robinson-Treitel Reader* was published, "compiled as a service to the industry by Seismograph Service Corporation, a subsidiary of Raytheon Company." Over the years *The Reader* went through several editions. In 2005 the SEG determined that the three top books ever published in applied geophysics were a dictionary, a popular text book, and *The Robinson-Treitel Reader*.

Chapter 3. Geophysical Analysis Group

Geophysical Analysis Group in 1953

By February 1953 fourteen oil and geophysical companies had joined to sponsor the GAG. Its Advisory Committee had one or two representatives from each sponsoring company. Some representatives were in geophysical research while the others were in geophysical operations.

The winter meeting of the GAG with the Advisory Committee was held in Dallas on January 30, 1953. Digital signal processing and deconvolution was akin to the invention of the telescope. When an invention is made one wants to use it on real data. Galileo looked heavenward and discovered the moons of Jupiter. We wanted to process geophysical data in order to get a clearer picture of the inside of the earth. I presented the computing scheme that MIT planned to use, including operator solutions to be done by Raytheon and the prediction-error process on Whirlwind. It was agreed that the initial test areas should include at least five shot holes in a profile where the record quality did not change appreciably from shot to shot. Then, several seismograms would be obtained from a single shot hole with variable depth and size of the shots. The Texas Company, Continental, and Magnolia each agreed to try to get a set of such seismograms together and sent to MIT. At least one of these three sets would be studied so that a fairly detailed report can be made by MIT in June 1953.

In addition to these sets, other types of sets were considered by the group. It was decided that sets of more complexity would then be supplied by other sponsors. Stanolind, Atlantic, and Cities Service each agreed to try to get suitable records released for this later study. I presented the deconvolution scheme that MIT planned to use. Initially, the GAG would plan to take about ten cases per seismogram. Since the capacity is estimated at 60 cases per month, this would mean six seismograms to be analyzed each month. The purpose was to verify the validity of deconvolution as a seismic processing scheme. The results were to appear in the next GAG reports, together with theoretical work on digital signal processing as applied to seismology.

The Advisory Committee gave the GAG four months to carry out this task. Early in June 1953, the GAG would prepare a report covering this work. The Advisory Committee would meet at MIT, at least ten days later, to discuss and examine the report. The Texas Company was the first to submit seismograms to the GAG for this purpose. As a result their records were used.

On February 2, 1953 the GAG held a meeting with Raytheon representatives. A few days later Raytheon sent prices for the work involved in the programming, the reading, the computation, the supplying of operator coefficients, and the final charting of the error curves (deconvolved traces) for the seismograph records to be supplied by the GAG. An agreement was reached with Raytheon on February 9, 1953 whereby the GAG would pay $2800 for programming and analysis and order a total of 320 full-case computations at $16.50 each, with the understanding that half-case computations could be prorated. Thus the contract came to a total of $8,080 made up of $2,800 and $5,280. There was an understanding that there would be flexibility so changes could be made underway. GAG did make changes which increased the cost.

In February 1953 Mark Smith and William Walsh joined the GAG as research assistants. Barbara Halpern became secretary and Irene Calnan became technical assistant. Simpson was finishing up his PhD dissertation so his available time was limited. I started teaching Smith and Walsh about digital signal processing and computer programming. Raytheon and I were left do the seismic deconvolution processing.

Raytheon decided to use a machine at the University of Toronto for the GAG computations. The Toronto machine was called FERUT (for Ferranti at UT). Raytheon selected this machine for this project after a careful survey of the few large scale electronic digital computers available for commercial computations.

Some of the advantageous aspects of the Ferranti Computer for seismogram analysis work were: a fairly satisfactory input speed, very satisfactory computational speed, and large drum storage. A

Chapter 3. Geophysical Analysis Group

disadvantage of the machine for this type of computation was its small electrostatic storage (RAM). For an extensive computation, such as the one under discussion, many instructions were required (some 1,000 instructions), as well as large amounts of data. When read in initially all this information was stored on the magnetic drum, and must later be transferred to the electrostatic storage for use in many successive relatively small blocks. The most important consequence of this fact was the difficulty of preparing the code and consequent possibilities of coding errors which must be located and corrected before the problem was run. Another limitation of the Ferranti Computer was its slow output speed. Despite many unforeseen circumstances due to breakdowns of the FERUT computer, Raytheon carried out its contract, which enabled us to prepared GAG Report No. 4 and No. 5.

On July 8, 1953 we submitted MIT GAG Report No. 3, *Case Study of Henderson County Seismic Record, Part I* to the Advisory Committee. This report contained computational results on Magnolia Petroleum Co. Record 10.9. The record was shot in Henderson County, Texas, where about sixty feet of loose Carrizo sand is underlain by Wilcox sandy clay with lignite stringers. Magnolia considered this a problem area because of the loose sand. Record 10.9 was an interference record that exhibited no clear reflections. However, Magnolia had marked five reflection times on the top trace. A different shooting procedure, which showed the reflections, provided the basis for indicating these refection times. Deconvolution easily picked up all the reflections on Record 10.9. Again the efficacy of deconvolution was demonstrated.

On July 21, 1953 we submitted MIT GAG Report No. 4, *Linear Operator Study of a Texas Company Seismic Profile, Part I* to the Advisory Committee. This report contained computational results on the records that The Texas Company sent to the GAG in February 1953 (Records 12.0 -12.18). These interference records had a highly ringing character. As a result, the autocorrelations were also highly ringing. We ran 30 deconvolutions on each of 10 seismic records. In other words we had run 300 deconvolutions in all. Each digitized trace and each deconvolved trace had about 600 points, which represents 1.5 seconds of seismic time at a spacing of 2.5 ms. In the 300 deconvolved traces there were

300 times 600, or 180,000 points to plot by hand. Corners were cut to get the job done in time. A seismic attribute of the deconvolved trace was plotted, not the trace itself. Each value of the attribute was based upon ten consecutive values. Only every tenth point of the attribute was computed and plotted, because the complete plot would require ten times the work. The results were good.

The summer meeting of the GAG was held on August 12-13, 1953. The representatives had planned to meet In June 1953. The reason for the two-month delay was due to the machine downtime on the FERUT. The representatives were rightly discouraged. However, the computing feat in itself was remarkable. No reflections at all had been marked on any of the records submitted by The Texas Company (Texaco). The study showed that the deconvolution removed the reverberations and yielded a consistent set of reflections. This achievement was a turning point.

The representatives entered into a lively discussion. All at once everyone became involved in one way or another in this venture. No one liked the way in which the deconvolved traces were graphed. This poor presentation was a drawback. The numerical values of the traces and of the deconvolved traces were given as backup material, but just the sight of this mass of numbers was daunting.

Attention turned to how the results should be displayed. There was considerable discussion. The committee suggested to the GAG that the deconvolved trace should always be plotted and never the attribute. A uniform plotting scale should be used and the traces should be grouped on a single large sheet for easier study. A suggestion was made that automatic plotting equipment be investigated. The GAG said it planned to use the oscilloscope screen on Whirlwind that was capable of displaying real-time text and graphics.

In September 1953 the GAG took on four new research assistants. They were David Bowker, Robert Bowman, Freeman Gilbert, and May Turyn. In February 1954 Sven Treitel also became a research assistant. During the fall of 1953, I undertook a training program for the new research assistants. This program consisted of regular conferences and discussion periods with the new members and the presentation of a course on

Chapter 3. Geophysical Analysis Group

digital signal processing with geophysical applications. Sven Treitel also participated in this course. Each research assistant was assigned a task. For example, I assigned, to the Robert Bowman, the task of deconvolving the ringing seismic records provided by the Atlantic Refining Company. A *ringing record* is one in which strong reverberations mask most of the primary reflections.

MASSACHUSETTS INSTITUTE OF TECHNOLOGY   Requisition   B   20764

ORDER FROM	Raytheon Manufacturing Co.		DATE Sept 28, 1953
ADDRESS	148 California Street Newton, Mass.		SHIP VIA
DELIVER GOODS TO E. A. Robinson 20E-222	CHARGE Geophysical Analysis 2490 DEPT.		SIGNED BY
QUANTITY	DESCRIPTION		
	Computing Services for the Geophysical Analysis Research Project, Dept. of Geology and Geophysics, MIT from Raytheon Mfg. Co. as follows:		
	Price of programming and coding		$2,800.00
90	Cases (with Operator Coefficients, Error Curves, and Variance Curves) at $16.50 each		1,485.00
60	Half Cases (Operator Coefficients only) at $8.25 each		495.00
200	Half Cases (Operator Coefficients only) at $22.00 each		4,400.00
GOODS RECEIVED	BILL APPROVED ON	Total price	9,180.00

SEND ORIGINAL REQUISITION WITH YELLOW CARBON COPY TO THE BURSAR

As of September 28, 1953 Raytheon had performed 90 full cases and 260 half cases of deconvolution on an assortment of seismic records provided by the oil companies. For the half cases, we used Whirlwind to do the numerical filtering to obtain the deconvolved traces. In all we had 350 cases of deconvolution on an assortment of seismic records. I had diligently studies the efficacy of these. All of the deconvolution filters revealed consistent sets of reflections, whether hidden or not. The computed reflections were accurate in every case in which we had knowledge of the true reflections as revealed by the company that had taken the records. This result served the basis for the seismic convolutional model.

Geophysical Analysis Group in 1954

On March 10, 1954 the GAG submitted MIT GAG Report No. 6, *Further Research on Linear Operators in Seismic Analysis*. This report represented work done since August 1953. The meeting of the GAG with the Advisory Committee took place at the Stanolind Research Center in Tulsa on March 29-30, 1954. We covered the material in GAG Report No. 6. One section demonstrated that a digital filter could do the same job as an analog electric filter in the classic example of a known signal superimposed in a noise seismogram.

The highlight was a section on the deconvolution of the ringing seismograms that the Atlantic Refining Company had supplied to the GAG. Deconvolution had removed the reverberations. The hidden reflections were revealed. H. F. Dunlap, the representative of the Atlantic Refining Company, said that the success of deconvolution on these ringing seismic records is an example of where deconvolution is a benefit. In previous meetings, we had always supplied the sponsors with a large number of deconvolution results. In order to focus attention, here we had picked one particular case, namely one in which deconvolution removed the reverberations.

There was a general reluctance to accept the possibility of digital computers ever replacing analog methods. There seemed to be a distrust of digital computers. They wanted the GAG to find the analog equivalent of deconvolution in terms of frequency filtering. Then deconvolution could be added as another feature of their seismic analog processing scheme. I knew that such an endeavor was futile. Digital was here to stay.

We had one more thing to offer in our attempt to convince the representatives that deconvolution could be commercially implemented. Clippinger and Levin had come to Tulsa and they made a strong case. They passed out the report *Utilization of Electronic Digital Computers in Analysis of Seismograms* by R. F. Clippinger, B. Dimsdale, and J. H. Levin. The report described the role of Raytheon in programming, coding, and computational tasks relating to the GAG work. It said that the Raytheon computer service was ready to perform deconvolution for the industry and gave prices for the service. Raytheon was also ready either to obtain or to build the digital computers that were most suitable for seismic work. In particular, Raytheon proposed using the new IBM 701 computer. Raytheon proposed to input the seismic data to the computer by magnetic tape. Again Raytheon tried to

Chapter 3. Geophysical Analysis Group

sell digital seismic processing to the oil companies. However, the companies were still not ready.

If the GAG had announced that we have found an assortment of analog frequency filters that could do deconvolution, the oil companies would have been delighted. Instead we said that we have found that digital time-domain filters can do everything that analog frequency filters can do. We even went one step further. We said that all of the analog methods could be done by digital signal processing. In fact, the digital way provided greater accuracy than the analog way. In effect, we proposed that the oil companies abandon their analog processing all together, and replace it with digital processing. Clippinger said that Raytheon was ready to obtain or build all the elements required for digital signal processing from input to output. It would be an upheaval, a digital revolution. The oil and geophysical company representatives were not ready to undertake digital seismic processing at this time.

There were was lengthy discussion at the second day, March 30, 1954, of this Tulsa meeting. The oil company representatives expressed their usual concern about the noise on a seismogram. They were not thinking about the ambient noise such as traffic noise and wind noise. Such noise is minor and it was of no concern in those days. The reflections on a trace made up the signal. All else was considered noise. The *additive model* says that the trace is the *sum* of signal and noise; that is,

$$\text{trace} = \text{signal} + \text{noise}$$

The object was to obtain a measure of the signal-to-noise ratio as a function of frequency. Analog filters could then be designed to pass frequencies in zones of high signal-to-noise ratio and to stop frequencies outside of these zones. I still have my handwritten notes. The underlying problem was that they could not understand deconvolution in terms of the additive model.

My mind raced back to October 1950, the first time I had ever seen a seismogram. Prof. Wadsworth was an expert in dealing with weather records. Prof. Hurley had remarked that the wiggly seismic traces looked much like weather records. The dialog between Prof. Wadsworth and Prof. Hurley initiated the events that led up to the GAG. However, weather records and seismic traces are not alike at all. Suppose you take a tracing of the temperature over a 24 hour day on Monday. The Friday tracing would not be the same as the Monday tracing. Suppose you take an explosion seismic trace on Monday. The subsurface of the earth is the same on Friday as it is on Monday. The Friday trace would be the same as the Monday trace. It does not matter what day of the week that you take an exploration seismogram. The noise on a seismic trace is signal-generated (i.e., explosion-generated) noise. It is only there because of the explosion, and it is not random but it is an expression of all the things happening to the travelling seismic waves in the unknown depths of the earth. The convolutional model takes care of these complications in an approximate way; the additive model does not.

The empirical evidence was that a causal deconvolution filter gives the knife-sharp impulses that represent the true reflections. This empirical evidence was provided by the 350 successful cases of deconvolution on an assortment of explosion seismic records. Now I had to provide a model. On April 2, 1954 back at MIT, I asked Irene Calnan, the GAG technical assistant, if she would like to do some work with me. All the cases of deconvolution done by Raytheon were two-channel. Irene and I ran a few cases of single-channel spiking deconvolution on Whirlwind. From one of these cases, I took the deconvolution operator (i.e., prediction error operator for unit prediction distance). The deconvolution operator was causal by design. It had to be stable because it only had a few coefficients. I had Irene calculate its causal inverse on the Marchant calculator. The causal inverse was infinity long. The important thing that happened was that the causal inverse damped out, which means that the causal inverse was stable. See Figure 9 on page 105 of this book. Because the causal deconvolution operator and its causal inverse were each stable, it followed that they were each minimum phase. Because the causal inverse is the seismic wavelet, we had numerical confirmation that the seismic wavelet is minimum phase. It was the first wavelet ever computed from empirical data. As usual, I wanted to compute a lot more examples. If I were to graduate in June, I only had a few weeks to finish my PhD thesis. Irene looked at me thoughtfully and said, "Enders. The wavelet is beautiful. Be satisfied." Irene's minimum-phase wavelet gave numerical confirmation of the seismic convolutional model. I wrote in Chapter VI, "Therefore it is not unreasonable to assume that this unique minimum phase wavelet is the wavelet generated by a physical phenomenon and the others are not."

On my thesis, Irene helped me with the figures and Barbara Halpern did a beautiful job of typing. There were many equations. However, at the last minute, I learned that I had to satisfy the academic requirement of attending the MIT undergraduate geology summer camp in Nova Scotia. There would be no degree in June. Before leaving for Nova Scotia, I happened to visit the basement stacks of the MIT library. I found a friend on his hands and knees scrubbing the floor with a bucket of soapy water and brush. He had just completed his PhD thesis in the Mathematics Department under Professor Norman Levinson. My friend explained that the Library had an offset press. To get the Library to print and bind his thesis, he said he would do anything. He advised me to do the same thing. As a result, I was able to get 200 copies of my thesis printed and bound as MIT GAG Report No. 7, *Predictive decomposition of time series with applications to seismic exploration*, dated July 12, 1954.

After Nova Scotia, I had to spend time on active duty in the Army in Virginia. The GAG Advisory Committee Meeting was held at MIT on September 14, 1954; it was the last meeting at which I spoke. At the meeting I presented the results of my thesis, namely the seismic convolutional model as the basis for deconvolution. The *convolutional model* says that the trace is the convolution of signal and noise; that is,

$$\text{trace} = \text{signal} * \text{noise}$$

which is the seismic case is

$$\text{trace} = \text{wavelet} * \text{impulses}$$

The trace is known. The wavelet and the impulses are unknown. The model gives us one equation with two unknowns. After my talk there was no discussion, but the Committee did have bound copies of MIT GAG Report No. 7 in their hands thanks to Prof. Levinson's student.

The day was saved by Prof. Simpson. He had written MIT GAG Report No. 8. It was based entirely on the additive model. It was good work and it was well received. There was a long and enthusiastic discussion.

Fifty four years after 1954, Professor Tad Ulrych of the University of British Columbia (*CSEG RECORDER*. April 2008, Vol. 33, No. 04) said: "I believe in fact, that in every field of endeavor, when people do inversion, at some point that inversion has roots in geophysics. Perhaps I am wrong, but as far as I am concerned, the first person to solve the problem of one equation with two unknowns, usefully, was Enders Robinson."

Geophysical Analysis Group in 1954-1957

In June 1954 Stephen Simpson had been promoted to the rank of Assistant Professor and became the director of the GAG. In October 1954, I went to work for Gulf Oil Company. Prof. Simpson then proceeded to change the research direction of the GAG. No longer would the "long and tedious calculations" be done to process and deconvolve actual geophysical signals. Not another deconvolution of a seismic trace was ever done by the GAG. The arrangement with Raytheon would end. The last deconvolutions ever done by the GAG were the single-channel cases that resulted in Irene's wavelet.

Instead the GAG started to do suitable things. The professors thought that the funds would be better used for acceptable academic research. Prof. Robert Shrock, head of the Department of Geology and Geophysics, later remarked, "It was not surprising that there was a question about the future of high-speed computers; even experienced scientists and engineers did not foresee what was to come." However, the research assistants on the GAG already had a large number of digital signal-processing projects underway. In completing these projects, these assistants produced a significant amount of new and useful results, a major accomplishment.

Prof. Simpson wrote this letter to Dr. Daniel Silverman, who was the chairman of the Advisory Committee of the GAG.
    March 21, 1955
    Dr. Daniel Silverman

Stanolind Oil and Gas Company
P.O. Box 591
Tulsa, Oklahoma

Dear Dr. Silverman:

I should like to ask your counsel on a budgetary matter which I intend to discuss at the April 1, 1955, business meeting. First let me outline what I have in mind and then question you about it. In this letter I will make the assumption that the Advisory Committee members are willing to sponsor our work for another year.

During the past year, the financial, as well as the directorial, responsibilities for G.A.G. have been turned over to me. Some of the former responsibilities are more difficult for me to evaluate than are the research ones; especially the question of justification of expenditures. This difficulty stems from my inadequate understanding of the feelings of the Advisory Committee members on these matters. Broadly, then, my problem is to get an expression of opinion from the Advisory Committee to guide me.

Specifically, however, I would like to request that I be permitted to use money from our 1956 funds for purposes which do not truly represent G.A.G. operating expenses. This money would be used mainly to assist the starting of new geophysical research and the progress of existing work in our department. I realize that the assistance of extra-G.A.G. research is not strictly within the scope of G.A.G. activities, but I feel that I am not entirely out of order in making this request for the following reason. Most of the people whose research would be benefited have given real service to G.A.G. administrative or otherwise, and this would be a way of remunerating them for their services.

On the other hand, I would not like to think of this money merely as compensation for direct or indirect services received against which it must be held accountable, since it would in reality be serving broader goodwill purposes. In this light I would include also certain interdepartmental expenses that might arise.

I have the feeling that the Advisory Committee members would agree in principle to some of these expenditures, but might question the amounts I have in mind, and perhaps this is the real problem to be settled. This completes the outline of my proposed request. Basically I would like to set aside an amount of money as a sort of "goodwill' fund, the expenditure of which would not have to be justified directly in terms of project expenses or salaries.

I should like now to ask your advice on this proposal. I would appreciate your considering these problems so that we could discuss them in New York prior to the G.A.G. meeting. First of all I should like to know of your own reactions, and secondly how you feel the Advisory Committee as a whole would react. I do not want to jeopardize our relations with the Committee members by making a request of this type, and if you feel that this danger is real I would be very glad to modify the request in accordance with any recommendations you might have. There is no need for you to write me in answer as I plan to get in touch with you in New York. As I recall, you plan to stay at the Statler during the S.E.G. meetings.

Sincerely,

Stephen M. Simpson, Jr.

On July 18, 1955, Prof. Simpson wrote:

July 18, 1955

Dr. Daniel Silverman
Stanolind Oil and Gas Co.
P.O. Box 591
Tulsa, Oklahoma

Dear Dr. Silverman:

Please pardon me for not writing sooner but I wanted to get the Renewals of Agreement out of the way first, and also to reach a decision about a proposal for A.P.I. support.

First of all I would like to summarize for you the Renewal status as of July 15

Atlantic	Renewed
Amerada	Out
Calif. Research	Renewed
Cities Service	Out
Continental	Out
Gulf	Renewed
Magnolia	Renewed
Phillips	Renewed
Stanolind	Renewed
Standard Oil Dev.	No reply
Sun Oil	Out
Texas Instr.	No definite answer (probably will renew)
United Geophysical	No reply
Ohio	No definite answer

As you must have seen from the carbon copies of letters sent you, M.I.T. is assessing the six companies who renewed for $4,000, subject to a partial refund.

July 18, 1955

My feelings about the future of GAG hinge largely about two things - one, the problem of administering research of this size, and two, a certain amount of indeterminacy in my own plans.

I have found that running GAG by myself has taken more of my time than it should in view of my other institute duties. At the present there does not appear to be a real possibility of my getting an assistant director next year. Hence, in whatever form GAG might continue in 1956-57, I would like it to be on a reduced scale.

It seems to me that a proposal on my part to A.P.I. at this time more or less obligates me to a two or three year project beginning a year from now - a commitment which I do not quite feel like making now.

This time next year the wave of four research assistants who came on in the fall of '54 will have had two years training (we are taking on only one new assistant this fall). Probably there will be about three students who will want to do thesis work in this field. I therefore visualize one final extension of GAG on its present basis to allow these students to produce.

I would think that I might be able to sell the companies on this extension on the basis of the investment training-wise that the group has in these students and on the fact that the budget would probably be about half its present size so we could get by with even fewer companies than we now have.

If it should develop that I can get administrative assistance, then perhaps we should modify our plans. Unfortunately very few people are around M.I.T. this summer, and I should take some action with regards A.P.I. quite soon. Hence I would very much appreciate an opinion from you as to the reasonableness of this approach to the problem of next years continuance. If you feel I should sound out the companies individually I can do that, but your own comments, at least in principle, on this course of action would be very helpful.

I thoroughly enjoyed my recent tour of the petroleum research labs, and my visit with your group. Unfortunately my schedule was so crowded I couldn't make the second visit as we had tentatively planned. I feel the trip will be of real value to me in directing this coming year's work.

Sincerely,

S. M. Simpson, Jr.

SMS|bh

On January 10, 1956, Professor Simpson wrote the following letter asking for the "dissolution of GAG" in six months' time.

January 10, 1956

Dr. Daniel Silverman

Stanolind Oil and Gas Co.

P. O. Box 591

Tulsa, Oklahoma

Dear Dr. Silverman:

I should like to present for your comments a proposal regarding GAG next year. From our point of view continuation of a large project devoted to linear operators (digital signal processing and deconvolution) in seismic analysis, is unattractive for several reasons. First, I feel that further discipline-oriented research of this type would not be as fruitful as it has been in the past. Furthermore a project of the present size might suffer from lack of adequate supervision.

What I propose, therefore, is dissolution of GAG as such as of June 30, 1956, with the project balance to be used by the Department to sponsor geophysical research under conditions to be mutually agreed upon.

These conditions could take the form of

1) Specification of field or fields of study

2) Specification of people under whose direction the money is to be spent.

3) Specification of some form of report on work so sponsored.

We would naturally desire to have these conditions as flexible as possible. A workable arrangement might be for the Department to select the staff men under whom the money is to be spent with each such man responsible for a report (research summary, article reprint, extended thesis abstract etc.) to be prepared and sent to the individuals who now comprise the Advisory Committee when his allotment is spent. More than likely I would continue to have several graduate students working under me so that at least a portion of this sum would be spent an extension of our seismic work already underway.

In any event, it would help us greatly in our thinking if you could let us know of your own reactions to the above proposal, your estimation of its reception by the Advisory Committee as a whole, as well as any additional comments or alternative proposals you might consider appropriate. I am looking forward to your counsel on these matters and, hope to hear from you shortly.

Sincerely,

S. M. Simpson, Jr

Chapter 3. Geophysical Analysis Group

Today observational sciences deal with digital signals. Cameras are digital; videos are digital; seismic is digital; audio is digital, telephones are digital; radio and television are digital. Digital signal processing has been fruitful over the years.

Freeman Gilbert and Sven Treitel were research assistants on the GAG at the time. Freeman Gilbert became a professor of Geophysics at the Institute of Geophysics and Planetary Physics (IGPP) at the Scripps Institution of Oceanography and the University of California, San Diego. He was elected a member of the National Academy of Science. In 2003, he wrote me:

> When I noticed your name in the notes for the MIT class of 1950, I had a few brief thoughts about the GAG. Little did we know what the next half-century would bring. I hope that you feel that you have taken part in, and in fact have been the primary leader of, a revolution in geophysical prospecting.

Geophysical Analysis Group 1954 and Beyond

Sven Treitel wrote *The MIT Geophysical Analysis Group (GAG), 1954 and Beyond* published in GEOPHYSICS 2005. Excerpts from the paper are given below.

ABSTRACT

> The MIT Geophysical Analysis Group (GAG) laid the groundwork for the so-called "digital revolution" in exploration seismology. An earlier article by Robinson (*The MIT Geophysical Analysis Group (GAG), from Inception to 1954,* GEOPHYSICS 2005) traces its history from its earliest days till 1954. Here the story continues with GAG's subsequent evolution until its end in 1957. But that was just the beginning: during the sixties and seventies the new digital technology spread throughout the oil and service industries worldwide, making it possible to develop progressively more sophisticated seismic processing and imaging algorithms that permanently changed the landscape of geophysical exploration.

INTRODUCTION

> I began graduate school in MIT's Department of Geology and Geophysics in September 1953. I was offered a teaching assistantship as a lab TA in mineralogy for that fall, and became an

official member of GAG at the beginning of the 1954 spring semester. It would be impressive to claim that I had already developed a passionate interest in geophysical time series analysis, yet nothing could be further from the truth: I simply went where those assistantships were offered. Before becoming an official GAG member, my advisor thought I should take an introductory time series course taught by Enders Robinson. ….

Once I joined GAG I was taken under the wing of Steve Simpson, who took over as director after Enders received his PhD in the spring of 1954. Steve taught me how to program the MIT Whirlwind Computer and my first assignment was to code a series of algorithms developed by the MIT statistician J.G. Bryan to test the hypothesis that a time series is stationary. …

GAG's TECHNICAL LEGACY

Today it is generally recognized that the GAG launched the digital revolution in exploration seismology. In fact, its efforts were among the very earliest successful applications of digital signal processing. Enders' preceding article presents a fine picture of its growing pains. GAG's high-water mark was reached on July 12, 1954 with the appearance of GAG Report No. 7 entitled "Predictive Decomposition of Time Series with Applications to Seismic Exploration," the full-length version of Enders' PhD thesis submitted to the Department of Geology and Geophysics.

The above model of a seismic time series, or trace, has continued to be one of the industry's workhorses to this day. All that has changed is that rather than speak of "predictive decomposition" we now speak of "predictive deconvolution." The thesis laid the groundwork for much of what was to come later; its six beautifully written and succinct chapters are as relevant today as they were fifty years ago. The figure below is taken from the June 1967 issue of GEOPHYSICS republication of Enders' thesis. It sketches the basics of what today is known as predictive deconvolution.

During GAG's three remaining years, significant research was undertaken in the design of linear least squares single and multi-channel operators in the presence of noise. The groundwork for the design of stable inverse filters was laid at that time, and the relationships between a digital filter's physical realizability and its stability were clarified. Much attention was given to the properties, origin, and treatment of different kinds of seismic noise. A clearer understanding of the fundamental differences between random noise and coherent noise led to early designs of least-squares multiple attenuation filters.

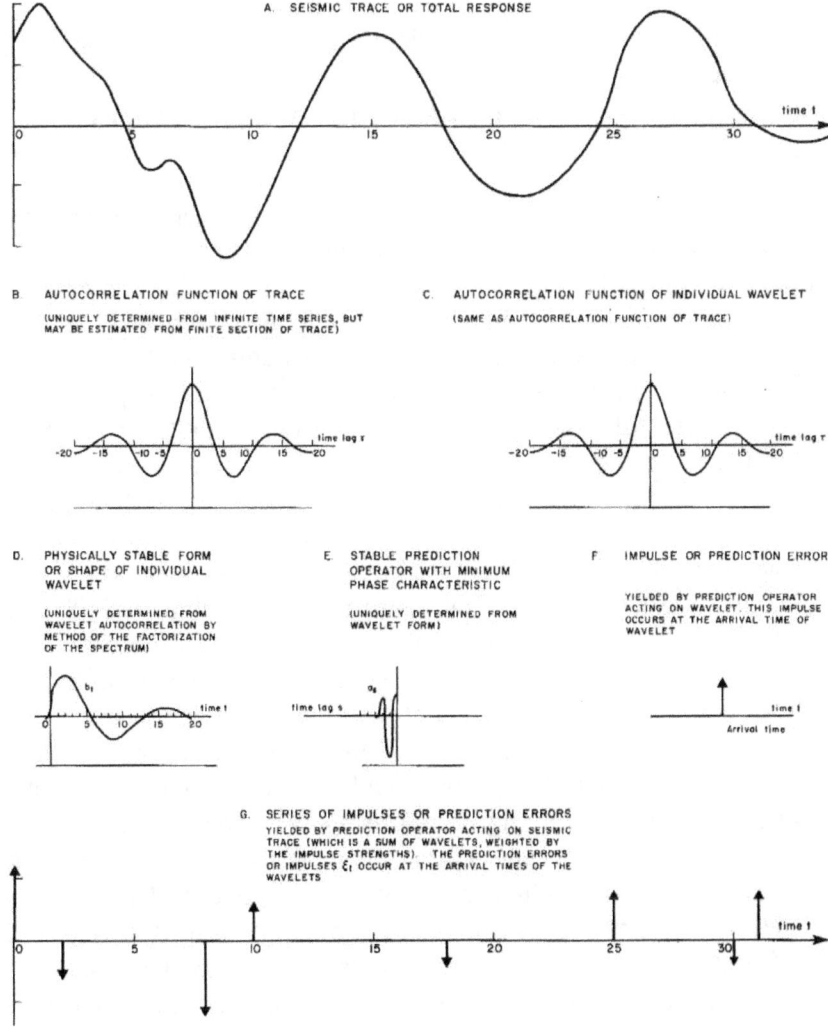

By early 1957, the project began losing steam. With twenty-twenty hindsight, too much attention had been given to the more arcane statistical aspects of time series analysis at the expense of further work with real exploration seismograms. At the same time, only a few of the sponsoring companies had people on their staffs with a good grasp of the technical issues GAG had begun to address, so the sorely needed guidance from industry was not forthcoming. Nobody in GAG then had any significant industrial experience. Several sponsors had already dropped out and by June of 1957 GAG was shut down. It is worth mentioning that this lesson has not been lost on later consortia in our industry: those which survive tend to be directed by individuals well aware of the importance of doing innovative science along with "sticking to the real data" and spinning off practical results.

GAG's LATER IMPACT: THE COLLABORATION BETWEEN ENDERS ROBINSON AND SVEN TREITEL

Upon receipt of my PhD from MIT in June 1958 I accepted a position in geophysical exploration in Havana, Cuba with what today is ChevronTexaco. Given that GAG came to an end in 1957 and a dissertation dealing with time series was not an option, and I ended up working with wave propagation in lossy media. [*On the dissipation of seismic energy from source to surface*. 175 pp, 1958, PhD Thesis, Prof. Gordon MacDonald, faculty advisor] In early 1959 Fidel Castro came to power and US oil companies were soon declared non-grata. A year later I joined the Tulsa Research Center of the Pan American Petroleum Corp., later known as Amoco. Dan Silverman was then the Director of Geophysical Research, having headed GAG's Advisory Committee throughout its lifetime. Following a futile struggle with the electrical DC response of a layered medium, I reminded Dan of the impressive amount of unfinished business GAG left behind, and its significant potential for an oil company. Since few people in science can work in isolation (and I am not one of them), I convinced Dan to invite Enders as a half-time consultant. At that time he was visiting at the University of Uppsala in Sweden and e-mail was still decades away. Following an initial visit by Enders to Tulsa, we began a joint project with the goal of adapting former GAG results to the needs of an oil company, as well as to develop FORTRAN based software for easy implementation of the signal processing theory. In so doing we were able to remedy a weakness in the GAG reports: writing was frequently at a level not understandable to most people working as exploration geophysicists. The plan was to write a series of internal reports at a sufficiently basic level as to serve as self-teaching tools. In this quest we were largely successful. Within a few years several of these reports were published in GEOPHYSICS and other signal-processing oriented journals; they became quite popular with our readers. At the same time we developed accompanying software, largely in the form of well documented FORTRAN subroutines. As time went by, these programs became the basis for Amoco's early seismic processing system; I am told (Paul Gutowski, personal communication) that remnants of these prehistoric subroutines can still be identified in BP's present processing system.

This first stage of our collaboration lasted some three years (1962-1965). A second period of joint research took us from 1974 through the mid-eighties. We communicated entirely via air mailed notes. The week-long intervals between receipt of each other's hand-written notes were actually a blessing since such delays enabled us to think before responding, a luxury not always available in this age of instant communication. I also recall that at one time an accountant at our downtown office called to say that he saw no reason why the company should spend airmail postage for our correspondence, instead he proposed surface mail via the slow boat to Sweden. I convinced him that just this once he should splurge. ...

CONCLUDING REMARKS

The MIT GAG set the pace for a research philosophy in our industry whose essential aspects have survived to this day, as evidenced by the ongoing existence of a significant number of innovative academic consortia. The GAG gave us an early appreciation of what can be accomplished by a combination of faculty members with vision and an enthusiastic group of students willing and eager to take off in novel directions. At a time when most major oil and service companies have abandoned their basic R&D efforts, we must look toward the academic consortia to provide the technology needed to explore for those increasingly hard-to-find new reservoirs.

Official citation for asteroid Svenders:

(54820) Svenders = 2001 NV1 Discovered 2001 July 11 by Joe Dellinger and William G. Dillon at the George Observatory, Brazos Bend State Park, Needville, Texas. Named by the **International Astronomical Union** in honor of Enders Robinson (1930–) and Sven Treitel (1929–). In 1952, Robinson became the first to ever perform signal processing on a general-purpose digital computer. Robinson and Treitel later co-authored a landmark series of papers that founded the modern field of applied geophysical signal analysis. Asteroid Svenders is about magnitude 17.0, and is a bright asteroid. The International Astronomical Union (IAU) unites national astronomical societies from around the world. It also acts as the internationally recognized authority for assigning designations to celestial bodies (stars, planets, asteroids, etc.)

The growth of digital signal processing

In October 1954, I went to work with Gulf Oil Corporation. They sent me to the oil fields of west Texas where I worked on seismic exploration crews. After that I went to their Pittsburg Research Center in Pennsylvania. I received this correspondence.

```
                                              March 2, 1955


      Professor Norbert Wiener
      2-155

      Dear Professor Wiener:

            This is to inform you that I have forwarded
      your note of February 28, 1955, addressed to Mr.
      Robinson, to him. He is with Gulf Research and
      Development in Pittsburgh right now but we expect
      to see him back here for a short while around the
      end of March.

                              Sincerely,


                              Stephen M. Simpson, Jr.

SS:pds

cc:  Dr. Enders A. Robinson
```

The result of Wiener's note was that in August 1955 I went back to the Mathematics Department at MIT as an instructor in mathematics. I went on to be an associate professor of mathematics at the University of Wisconsin-Madison. In 1960 the Wisconsin Alumni Research Foundation gave me a fellowship to spend time at Uppsala University in Sweden with Professor Herman Wold. Digital signal processing was taking root. Computers were getting better. In the 1960s, the seismic industry converted entirely to digital signal processing. New oil fields, such as ones in Alaska, the Gulf of Mexico, and the North Sea, were discovered by the use of digital signal processing and deconvolution.

Chapter 3. Geophysical Analysis Group

Texas Instruments
INCORPORATED

6000 LEMMON AVENUE, DALLAS 9, TEXAS
DIXON-1781 • BOX 7045, LOVE FIELD • CABLE TEXINS

16 February 1953

Dr. P. M. Hurley
Department of Geology and Geophysics
Massachusetts Institute of Technology
Cambridge 39, Massachusetts

Dear Pat:

I am returning a signed copy of the agreement covering the Geophysical Analysis Research Program. You will note that I have changed the name from Geophysical Service Inc. to Texas Instruments Incorporated inasmuch as the latter corporate group is the one actually involved.

I have not included the check for $4,000 inasmuch as I rather gathered from the agreement that the contribution was subsequent to the signing of the agreement. If it is due immediately, will you please so advise me, or, if not, when payment should be made.

Much obliged!

Sincerely yours,

P. E. Haggerty
Executive Vice-President

THE MASSACHUSETTS INSTITUTE OF TECHNOLOGY

By _J. A. Stratton_
Vice President, _Feb. 6._ 195_3_.

Accepted and Agreed to

This _25th_ day of _February_, 1953_

TEXAS INSTRUMENTS INCORPORATED
(Name of Company)

By _P. E. Haggerty_

Chapter 4. MIT Reports

MIT Reports on Research

Reports on Research

MASSACHUSETTS INSTITUTE OF TECHNOLOGY   FEBRUARY, 1953   VOLUME 4, NUMBER 4

MIT published a monthly periodical entitled *Reports on Research*. Interestingly in the issue of February 1953 (Volume 4, Number 4) the article *Oil from mathematics* was published on finding oil by digital signal processing (deconvolution, then called prediction error filtering).

Oil from Mathematics

> Methods of statistical analysis may yield important conclusions regarding the earth's structure, and could be significant in increasing the world's yield of petroleum.

When MIT wrote this report, did they anticipate that not only would mathematics be significant in increasing the world's yield of petroleum, but it would be responsible for the discovery of amounts of petroleum that exceeded all expectations by orders of magnitude? The phase "oil from mathematics" was prophetic. Mathematics resulted in the conversion of the entire oil exploration industry from analog to digital in the 1960s. It was the first industry to undergo a digital revolution. As a result, vast new regions of the world became open to oil exploration, not only in extent but in depth into the earth. Exploration at sea became possible. Digital signal processing makes possible the discovery of the great oil and gas reserves that are recoverable by hydraulic fracturing. In Shakespeare, Jack Falstaff says, "Banish plump Jack, and banish all the world." Banish mathematics, and banish all the oil.

In the same issue the article *Sensing of magnetic cores* described the magnetic core memory developed at the MIT Digital Computer Laboratory. This research resulted in the conversion of the entire computer industry from unreliable memory to the highly reliable magnetic core memory. Whirlwind became the first dependable computer; all previous computers could not be trusted for unswerving continuous usage. Magnetic core memory was the breakthrough that made the computer industry possible. Whirlwind did not fit the accepted model of a general purpose digital computer. It was in effect a special purpose computer designed for one purpose; namely, the rapid assembly and transfer of digital signals. Take Whirlwind a few steps further, and the result is a special purpose computer designed for one purpose; namely, rapid digital signal processing. In other words, Whirlwind may arguably be called the prototype of the DSP chip.

Oil from mathematics (MIT, 1953)

We now give the text of *Oil from Mathematics*.

Methods of statistical analysis may yield important conclusions regarding the earth's structure, and could be significant in increasing the world's yield of petroleum.

The increasing amount of petroleum to serve the world's industrial needs has already caused many oil fields to become exhausted, and has made the work of the prospector increasingly difficult. Temporary relief from the constantly pressing needs for more oil was afforded, about two decades ago, when the reflection seismic method of petroleum prospecting was put into general use. In this technique, a charge of dynamite is exploded under controlled conditions and the resulting vibrations at different points on the earth's surface are recorded on graphs known as seismograms. The analysis of such seismic records, for a given disturbance in the earth's crust, yields valuable information of the structure of the earth's outer layer, and such information is of considerable economic value in increasing the likelihood of locating and operating new oil fields successfully.

In the past few years the search for new oil has required the investigation of areas in which petroleum reserves are increasingly difficult to locate. Present-day techniques, in which the geophysicist examines seismograms visually, are reaching the limit of effectiveness for these areas. If new oil fields are to be discovered and worked economically, improvements in prospecting techniques must be continually introduced. Perhaps the greatest step forward, in the immediate future, may be achieved by more penetrating methods of analyzing seismograms, through the use of statistical techniques in which modern high-speed computers may play a significant role. The application of modem statistical methods makes it possible to separate the desired energy reflections from extraneous variations on seismic records, so that more of the information obtained in the field may be put to practical use.

Research on this problem is being carried out in the Department of Geology and Geophysics at the Institute under the direction of a staff committee, with Enders A. Robinson in charge of operations. The project was initiated by Professor George P. Wadsworth, of the Department of Mathematics, who had previously applied statistical methods, utilizing so-called linear operators, to the analysis of meteorological and oceanographic data. The first person to realize the physical implication of linear operators and appreciate their applicability to the problem of single and multiple filtering and prediction, was Professor Norbert Wiener, of the Department of Mathematics. During World War II, he applied the method of linear operators to the solution of basic problems in radar systems, and in anti-aircraft fire control. Basically the new methods make it possible to bring into sharp focus the dynamic elements of a physical situation, by means of certain types of averaging processes which simultaneously suppress disturbing elements.

By applying appropriate averaging techniques to different sections of reflection seismograms, it was found that the dynamic characteristics remained fairly consistent except during intervals corresponding to a reflection. The inconsistency, or error, introduced gives a measure of the change in the dynamic properties, and hence may be used as an indication of the amount of energy reflected. By examining an error plot, derived from the original seismic record, one is able to pick off the arrival of reflected energy at places of high error. More directly, one may correlate places of high error with interfaces between sub-surface layers. Thus, such an error plot becomes a more powerful tool than the

Chapter 4. MIT Reports

original seismogram in determining the underground structure of the earth, and consequently increases the likelihood of success before actually drilling a well.

For the seismograms thus far analyzed, results to date have indicated that the new method of analysis yields valid error plots. Moreover, such a method may even differentiate zones of greater scattering which have no single major reflecting surface. In addition, the work has yielded valuable information on the dynamic characteristics of recorded earth movements, and may also form the basis of a considerable amount of theoretical work of value to geologists.

An open invitation to petroleum companies to collaborate in this research program has resulted in both financial support and close scientific liaison between the research group at M.I.T. and the research organizations of petroleum companies.

Making Electrons Count

In 1953, MIT produced the film "Making Electrons Count," which was sponsored by the Office of Naval Research. The film provides a fascinating tour of the Whirlwind computer facilities at MIT. It illustrates daily routines, problem-shooting and step-by-step procedures that computer programmers and other users go through at the research center. When the film came to naming some of the projects at MIT that used Whirlwind, the first project described was the GAG. The narrator in the film says,

> "The film which are about to see first shows a few examples of the types of problems in which computers can be useful, and then describes the efforts of a typical user in programming a problem for Whirlwind. Whirlwind has been involved in more than a hundred such computational problems, originating in many different departments of MIT. Take the Geology Department, for example. Seismic methods of prospecting for oil may seem a little strange to the onlooker. A charge is exploded at one point, and the sound reflected from various underground layers of rock is recorded in a number of other points. A great deal of information about underground formations can be determined from the sound

patterns, but only **after long and tedious computations** have been performed on them."

We may forgive the narrator for saying "long and tedious computations." The reason is that our digital signal processing used more computer time by far than other academic project on Whirlwind. Indeed, to the Whirlwind staff, our computations were not only long and tedious, but they also required a tremendous amount of input and output of seismic data. Digitized seismic data was a new idea. The quantity of seismic data involved was unending, because each oil company kept submitting an ever increasing supply of seismic records, many more that we could ever handle with the facilities at hand. Also Whirlwind was a military computer, with some time made available to the MIT staff for research. It was clear that the oil industry would have to make use of commercially available computers.

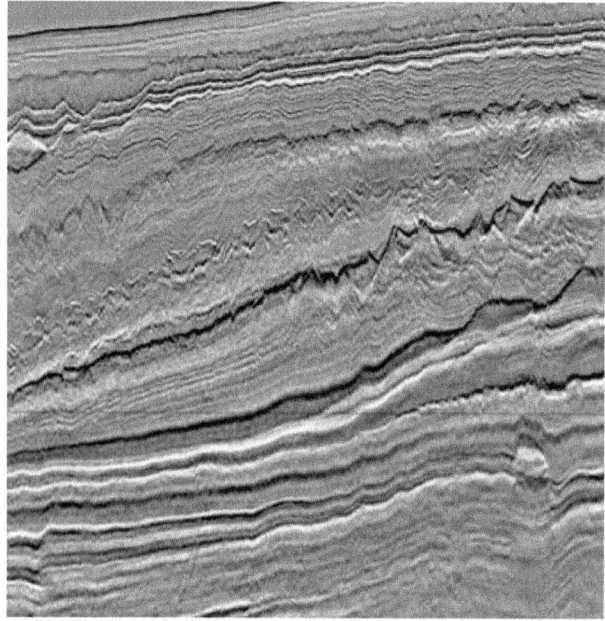

Figure. A high-detail cross-section of a 3-D image of the earth's subsurface obtained in the twenty-first century by means of the "long and tedious computations" used in digital signal processing and deconvolution.

Chapter 5. Raytheon Computer Services

IEEE Computer Pioneer Award

The IEEE Computer Society presents its Computer Pioneer Award to a person who made significant contributions to concepts and developments in the electronic computer field which have clearly advanced the state of the art in computing. The contributions must have taken place fifteen or more years earlier. In 1996 Richard Clippinger received the Computer Pioneer Award for his conversion of the ENIAC to a stored program computer. The IEEE Computer Society gave this citation.

> Richard F. Clippinger. Born 1913, East Liberty, Ohio; computing laboratory staff member, Aberdeen Proving Ground, who converted the ENIAC to a stored program computer using its read-only hand-set function tables.
>
> Education: PhD, mathematics, Harvard University, 1940.
>
> Professional Experience: ballistic research laboratory, Aberdeen Proving Ground, 1944-1952; Raytheon Computer Laboratory (later Datamatic Corporation, and later still EDP Division of Honeywell), 1952-1976.
>
> Clippinger went to the Ballistic Research Laboratory at Aberdeen Proving Ground in 1944. There he invented and developed the closed-chamber firing range, which rivaled the wind tunnel for measuring forces on a supersonic model. At Aberdeen he also worked in the development of numeric methods for solving ordinary and partial differential equations on the ENIAC, EDVAC, and ORDVAC. In 1952 he joined the Raytheon Computer Laboratory, which became Datamatic Corporation in 1954 and the EDP division of Honeywell in 1956. He was in charge of software development for the Honeywell 800 family until 1959 when he supervised the development of the FACT business language compiler by Computer Sciences Corporation.

Let us explain. As originally built the ENIAC did not store programs in memory as a modern computer does. Instead the programming was done by rewiring the physical components for each new problem. To solve a mathematical problem, components of the ENIAC would have to be assembled, mainly through wiring, into a special purpose computer. In a few minutes the machine ground out the required answers for this particular mathematical problem. The next mathematical problem would require a complete new rewiring from scratch.

Dr. Richard F. Clippinger was the head of the Mathematics Division at Aberdeen. In 1947 he set out to modify the ENIAC into a stored program computer. John von Neumann was a consultant to Aberdeen Proving Ground. Clippinger and Dimsdale would meet with him and they would fill him in on the modifications they were making on the ENIAC. By the summer of 1948 Clippinger had succeeded in converting the ENIAC into the world's first stored-program computer with a programming language. The programs were fed into the machine on the function tables (the banks of switch-controlled resistor matrices originally designed to hold input data).

The new method worked beautifully, but others wanted to go back to the original way. They said the ENIAC was at least six times faster with the old way. But Dimsdale came to the rescue. He said although that is true, it is not important. The old way requires some months to configure the ENIAC, and then a few minutes to do the calculations. The new way requires a few days to write the code, and a few hours to do the calculations. The old way of rewiring the machine for each problem was never used again.

Dr. Dimsdale wrote extremely efficient codes for deconvolution for the Geophysical Analysis Group. In fact, Dr. Dimsdale wrote a DSP code that was the first code ever to work the first time that it was run on the Ferranti computer with no corrections needed. Later Dr. Bernard Dimsdale left Raytheon and joined IBM. He became an IBM Fellow, the highest title for a scientist at IBM. At the Washington DC IBM Space Center, Dimsdale managed the operating system underlying TAGIS – which was the Targeting, Acquisition and Ground Information System for NASA. This operating system enabled the stations around the world to follow the space vehicle containing the NASA astronauts.

For years the ENIAC was the only computer reliable enough to solve large scale problems. None of these problems involved digital signal processing. The problems were usually of the type that involved finding solutions to specific differential or integral equations. Once the solution of one problem was obtained, the ENIAC would be turned to solve a completely different equation. A general purpose computer is good for solving a lot of different problems. However, for the solution of only one type of problem, a special purpose computer would be faster and easier to use.

In digital signal processing, just one type of mathematical problem is solved. In such a case, a special purpose computer is a good idea. Such a computer could be designed to do the specific tasks involved.

Whirlwind was a general purpose computer and it could perform deconvolution. But with its short word length and the required double precision arithmetic, Whirlwind was far from ideal in performing the massive number of multiplications required in a short period of time. A special purpose computer with long word length would be much better.

Of course, the GAG did not have the means to build a special purpose computer. However, the oil companies easily had the means. In 1953 Raytheon had access to the British Ferranti Mark 1 computer (which was the commercial version of the Manchester Mark 1 computer for which Alan Turing played a key role). The Mark 1 computer in question was installed at the University of Toronto to help in the design of the St. Lawrence Seaway. The Mark 1 had a long word length. It was much more suitable than Whirlwind for computing the coefficients of the deconvolution filters.

In place of a special purpose computer, the GAG would do the next best thing. It would use the Mark 1 to compute the coefficients of the deconvolution filters, and then use Whirlwind to carry out the filtering operations. In effect, the plan was to use two different general purpose digital computers in order to do digital signal processing efficiently.

For the oil companies, Raytheon would do the analog-to-digital conversion, deconvolve the signals, and plot the results. Raytheon had designs to build a special digital signal processing (DSP) digital computer for this purpose. It would have been the ancestor of the present-day DSP chip.

Discussion of machine solution of seismogram analysis problem

On August 4, 1953 we submitted MIT GAG Report No. 5, *On the Theory and Practice of Linear Operators in Seismic Analysis* to the Advisory Committee. This report contained research work done by the Geophysical Analysis Group since February 1953. Section 7 was *Discussion of machine solution of seismogram analysis program* by R. F. Clippinger and J. H. Levin of the Computer Services Section, Raytheon Manufacturing Co. We reproduce Section 7 in in the following seven subsections.

1. General description of computer on which the problem is solved

The machine used in the calculation is the Ferranti Computer, commonly known as "Ferut," constructed by Ferranti Limited in England and located at the University of Toronto. This machine was selected for this project after a careful survey of the few large scale electronic digital computers available for commercial computations. The Ferut operates in the binary number system, and has the following characteristics:

> Overall dimensions — Two bays each 16 feet long, 8 feet high, and 4 feet wide, and a control desk.
>
> Power consumed — 27 kilowatts.
>
> Number of components — About 4,000 vacuum tubes, 15,000 resistors.

Speed

> a. Multiplication time — 2.2 ms (i.e., about 450 multiplications per second).
>
> b. Addition time — 1.2 ms (i.e., about 833 additions per second).
>
> c. Input — Punched teleprinter tape read photoelectrically and fed in at rate up to 200 characters per second.
>
> d. Output — Punched teleprinter tape at 10 characters per second or direct printing by teleprinter at 6 characters per second.

Storage

> a. Internal high speed electrostatic storage of 256 words (a word being 40 binary digits, i.e., about 12.1 decimal digits).
>
> b. Magnetic Drum - storage facilities for about 16,000 words.

(1) Arithmetic operations within the machine can only be carried out on numbers in the electrostatic storage. Information stored on the magnetic drum must be transferred to the electrostatic storage before it can be used for computation. Once such a transfer is made, however, the information still remains on the drum and will remain there indefinitely until replaced by other information.

(2) In carrying out any computation, the machine must store its own instructions as well as the data required, and as pointed out above these are only accessible for use after being transferred to the electrostatic storage. This transfer itself is accomplished by means of one of the instructions. These facts are important from the point of view of discussing the advantages and limitations of the machine.

2. Aspects of the Ferranti machine.

Some of the advantageous aspects of the Ferranti Computer for seismogram analysis work are: a fairly satisfactory input speed, very satisfactory computational speed, and large drum storage. A disadvantage of the machine for this type of computation is its small electrostatic storage. For an extensive computation, such as the one under discussion, many instructions are required (some 1,000 instructions), as well as large amounts of data. When read in initially all this information is stored on the magnetic drum, and must later be transferred to the electrostatic storage for use in many successive relatively small blocks. The most important consequence of this fact is the difficulty of preparing the code and consequent possibilities of coding errors which must be located and corrected before the problem is run. Another limitation of the machine which is expected to be relieved sometime in the not too distant future is its slow output speed. While it is possible to reduce loss of time due to this cause by taking advantage of the fact that the machine can compute during printing, in any specific case, other coding considerations sometimes limit the extent to which this can be done. The above shortcomings have been largely overcome in the code which Raytheon has prepared.

3. Raytheon Computer Services Section Staff

Senior Staff

>Dr. R. F. Clippinger – Head of Computer Services Section. Before entering Raytheon, he was assistant chief of the Computing

Laboratory at the Ballistic Research Laboratories at Aberdeen Proving Ground, Maryland. In this capacity he was responsible for the operation of ENIAC, EDVAC, and ORDVAC in the solution of problems reaching the Computing Laboratory.

Dr. B. Dimsdale was formerly head of the numerical Analysis Branch of the aforementioned Computing Laboratory. As such he was responsible for analyzing the many varied problems received by BRL and for carrying out research directed toward development of numerical methods better adapted to modern computing machines.

Dr. J. H. Levin was formerly in charge of operations on the SEAC (National Bureau of Standards - Eastern Automatic Computer). Previous to that he had directed the activities of the IBM punched card installation at NBS and of the IBM and differential analyzer installations at the Ballistic Research Laboratories.

Figure. The ENIAC as it was in 1946 showing the wires which had to be rearranged in order to program the computer. In 1947-1948 R. F. Clippinger converted the ENIAC into the world's first stored-program computer. The program was stored in ROM, so the wires were no longer needed.

Chapter 5. Raytheon Computer Services

Junior Staff. The senior staff is assisted at present by a small number of able programmers with substantial experience in preparing and running problems for RAYDAC, MIDAC, the Ferranti Computer, EDVAC, etc.

Raytheon's Computer Services Section has active plans to expand, both on the analytical and programming levels, in accordance with the expanding demand for its services. A number of persons are in stand-by readiness to join the staff when called. In addition, arrangements have been made to secure outside consultation services during periods of peak load.

4. Prospects of price reductions - short range

It is anticipated that future production costs will undergo substantial reductions. Clerical costs for tape preparation (a rather intricate procedure requiring extreme care to eliminate any possibility of error) have been reduced as personnel became more experienced, and are expected to undergo still further reductions. Supervisory costs have similarly undergone reductions, and may also be expected to decrease further.

In addition to reduction of labor costs, there exist numerous possibilities for reduction of machine time per case. At present 5 minutes of machine time are required for a half case and 15 minutes for a full case. Only a relatively small part of this time is used in actual computing. For full cases, for example, computing takes about 5 minutes, and printing about 10 minutes. This situation is expected to improve with the introduction of a parallel high speed printer in the not too distant future. This will decrease printing time to 25 per cent of its present figure, or better. At present the cost of output printing for a full case is approximately $ 18. At the current rental rate this could be expected to be reduced to $4.50, or less, a reduction of at least $13.50. If one assumes that the improved facility results in increasing the rental by as much as $ 20 per hour, the printing cost would still be reduced to $ 5.40, or less.

A number of possibilities exist for reducing cost of solution on Ferut by amending the program. For example, the program may be modified so that the computer will take in the master tapes corresponding to the traces, and for each case will call for parameters and other identification information from a short supervisory tape. The proper portions of the traces will then be selected, and the computations carried out. When all required computations are completed, a new set of traces would be called in. This type of revision would eliminate the necessity of preparing specific data input tapes for each case. This is only one of several kinds of coding improvements that would result in cost reductions.

5. Current procedure

Three fundamental sets of quantities must be computed in this problem: the operator coefficients, error curves and variance curves. A computation involving all of these is called a full case and one involving only the operator coefficients is called a half case, although time-wise and cost-wise it is less than half of a full case. The first step for either the full or half case calculation consists of reading and recording of traces and preparation of tapes. The traces are read at intervals of $h = 2.5$ ms, and there are usually between 400 and 800 readings per trace. The readings for each trace are then punched on teletype tape, the tape is printed, the printed copy is compared with the original record, and any errors are corrected. Such a tape called a "master tape" is prepared for each trace, and is used to prepare later so called "partial tapes," and then "final data input tapes" for specific half or full cases.

These data input tapes for specific cases must incorporate not only the data from the traces but also certain parameters, which vary from case to case, including:

k-value of prediction distance (two values of k per case have been typical).

$M + 1$ - number of operator coefficients corresponding to one trace.($M + 1 = 4$ in all cases to date).

N - index of the first term in the operator time interval, or equivalent.

n - number of terms in the operator time interval.

J - number of traces used in the determination of coefficients and in the predictions. This determines the size of the matrix to be inverted.

Chapter 5. Raytheon Computer Services

Additional parameters for full cases are:

Range to be covered by error and variance curves.

The data input tapes consist of the appropriate segments of the master tape together with the parameter information. Still other information must be included on the tapes to tell the machine how many words are forthcoming and where they are to be stored. The information must be fed to the machine in blocks not exceeding a specified size (this is due basically to the way in which information is transferred between the electrostatic storage and the drum). These data input tapes must be checked and errors corrected. They are then sent to the University of Toronto for running and the printed results are returned.

The program carried out by the machine may be broken down into a number of principal parts as follows:

1. Read in parameters and data.

2. Form normal matrix.

3. Invert matrix using initial trial scale factor, ℓ, empirically selected using information about scale factor from earlier cases. Occasionally, more than one trial is necessary before an appropriate scale factor is found.

4. Compute operator coefficients for each value of k.

5. Substitute back in original normal equations. Print out indication of how well the equations are satisfied.

6. Print operator coefficients for each value of k.

7. Compute and print I_m and I_0 for each value of k.

8. Compute and print

$$\frac{1}{10}\sum(x - \hat{x})^2 \quad and \quad \frac{1}{10}\sum(x - \bar{x})^2$$

for non-overlapping intervals of 10 covering prescribed range of error curve. This step is performed for full cases only.

There are a number of checks on the machine computations. Some types of machine errors may cause the machine to go into a completely nonsensical sequence and it becomes immediately obvious to the operator that something has gone wrong. He can then take appropriate action. Barring this type of error, after the computer obtains the operator coefficients, it substitutes them automatically back in the normal equations and prints out an indication of how well these equations are satisfied. Also since we have a fair idea of what the final results should look like, it is possible to detect certain types of errors by inspection of the results. For example, we should expect to have $I_m < I_0$. If this is not so, then the case would be rerun. If the final results pass the machine checks and visual checks, they are accepted and submitted for graphing.

Except for certain instances where it is necessary to change the scale factor, the program described above is completely automatic, i, e. , no human intervention is required from the time the data is introduced into the machine until the time the final results are printed out. In the case of change of scale factor, under certain circumstances, this too is automatic.

6. Costs

Twenty-four man months of effort were required for programming and coding this problem and checking it in on the machine. The following tabulation shows the break-down of preparation costs.

Mathematical analysis, programming, coding, and checking into computer	$ 27,000
Computer rental for checking	2,600
Equipment	1,000
Total	$30,600

The foregoing represents Raytheon's actual investment in preparing the basic code for seismogram analysis work. Further expenditures of this nature are contemplated for improving the basic code which will result in reducing costs of computation on a production basis.

Estimated prices for computing full cases are based on the schedule given below. It is assumed that each case has 50 points in the operator time interval and 2 prediction constants, that 600 points are used from each trace, and that the linear operator depends on 3 traces.

Reading traces, recording, punching, checking	$ 20.00 per trace of 600 points
Preparation of partial tapes	12.00 per group of 3 traces
Assembly and checking of complete data input tape	26.00 per case
Computer rental	27.50 per case
Graphing of results	12.50 per case
Supervision, coordination	24.00 per case

The total price (in dollars) per case for a lot of N cases may be expressed by the formula

$$20T + 6P + 90N$$

where

T = number of traces on which the N cases are based,

P = number of partial tapes to be prepared.

Thus, for a lot of 100 full cases based on 36 traces in total, and requiring 24 partial tapes to be prepared, the price would be $10,008, or $100.08 per case.

7. Prospects of long range improvements

From a long range point of view it might be worth-while to seriously consider the possibility of doing this problem on a machine having magnetic tape input and having digital to analog facilities for output. It is understood that seismogram recordings are already made in some cases on magnetic tape. With the type of equipment under consideration it should be possible to feed the recordings directly into the machine and to program the machine for analyzing these recordings. The output would be converted from digital to analog form and recording pens would draw the error and variance curves directly. There is good reason to believe that a long range program of development along these lines might result in prices below $20 per full case.

Chapter 6. MIT PhD thesis 1954

Letter from Professor Herman Wold

On October 7, 1954 I sent a copy of my dissertation (GAG Report No. 7) to Professor Herman Wold in Uppsala, Sweden. He was one of the early pioneers of time series analysis. His reply sent on October 18, 1954 was forwarded to me in Lamesa, Texas. He wrote to me as follows: "First of all I wish to congratulate you on your thesis and excellent work with several interesting results and equally distinguished by the thorough treatment and clarity of exposition. You may imagine that I am happy to see my theorem of predictive decomposition subjected to such a brilliant review in the light of the subsequent development. And I am touched in seeing with what scrupulous care you have given reference to priority to results in my thesis that were novel at the time. You kindly invite me to give criticism and comments, and offer to submit my remarks to those who are receiving your thesis. Actually I have little or nothing to criticize. My predictive decomposition was primarily intended as an existence theorem, the emphasis being upon the distinction between the regular and the singular component, and the interpretation of earlier approaches as special cases of the stationary process. I was aware that there was a close relationship between the decomposition and the properties of the spectral function, but I did not enter upon this problem, for the simple reason that I did not master the general methods of spectral theory. The applications (the method of deconvolution) in your Chapter 6 are very impressive. I hope the contact now established will continue, and in particular I should like to hear your reaction on the above comments. Further I hope to make your personal acquaintance someday. Is there a chance that you might come over to Europe during the near future? In such a case, please do not forget you have a friend in Uppsala." This letter was the start of a life-long friendship with Professor Wold.

The content of Chapter VI of my PhD thesis is published in Geophysics, vol. 22, pp. 767-778, 1957. The entire thesis is in Geophysics, vol. 32, pp. 418-484, 1967. Chapter VI of my thesis follows in the next section.

Picture: Robinson (left), Wold (middle)

Chapter VI of MIT PhD thesis 1954

[EXPLANATORY MATERAL WRITTEN IN 2014 IS SHOWN IN SQUARE BRACKETS.]

The first section of Chapter VI describes Ricker's wavelet theory of seismogram structure. The second section follows:

Thus the seismogram may be visualized as the totality of responses to impulses, each impulse ϵ_t being associated with a disturbance which has traveled a certain path by refractions and reflections. These responses, or response functions, are the seismic wavelets b_t. The analysis of a seismogram consists in breaking down this elaborate wavelet complex into its component wavelets. In particular we desire the arrival times of the theoretical sharp impulses which produce these wavelets or responses.

There are two basic approaches which one may use toward the solution of this problem: the deterministic approach and the probabilistic or statistical approach. In the deterministic approach one utilizes basic physical laws, for example, in order to determine the shape of the wavelet, or the absorption spectrum of the earth. At all stages in such an investigation, one may compare mathematical results with direct and indirect observations of the physical phenomenon.

In this thesis we are concerned with the statistical approach. Such an approach in no way conflicts with the deterministic approach, although each approach has certain advantages and disadvantages which do not necessarily coincide. The emphasis we place on the probabilistic approach is due to its being the subject of investigation of this thesis. In practice the two approaches may be utilized in such a manner so as to complement each other.

Let us apply the probabilistic approach to one specific problem. Let us set up a hypothetical situation. Let us assume that a given section of seismic trace is additively composed of seismic wavelets, where each wavelet has the same shape or form. We shall assume that the shape of the wavelet is minimum-phase, that is, the discrete representation of the wavelet shape is a solution of a stable difference equation. Further, we assume that from knowledge of the arrival time of one wavelet we cannot predict the arrival time of another wavelet; and, we assume that from knowledge of the strength of one wavelet we cannot predict the strength of another wavelet. Finally, let us assume that the seismic trace is an automatic volume control (AVC) recording so that the strengths of these wavelets have a constant standard deviation (or variance) with time.

The specific problem which we wish to consider is the following: given the seismic trace described in the above paragraph, determine the arrival times and strengths of the seismic wavelets, and determine the basic wavelet shape. We shall discuss a theoretical solution of this problem, and shall also discuss a practical solution which involves statistical estimation.

Let us translate our assumptions about the seismic trace into mathematical notation for discrete time t. First let the shape of the fundamental seismic wavelet be given by the discrete minimum-phase time function b_t where $b_t = 0$ for t less than zero. In other words, the

Chapter 6. MIT PhD thesis 1954

value b_0 is the initial (nonzero) amplitude of the wavelet. Discrete minimum-phase time functions are discussed in Section 2.6 of this thesis (MIT, 1954).

Let the strength, or weighting factor, of the wavelet which arrives at time t be given by ϵ_t. That is, ϵ_t is a weighting factor which weights the entire wavelet whose arrival time is time t. The variable ϵ_t is the theoretical knife-sharp impulse of which the particular wavelet (i.e. the one which arrives at time t) is the response. For example, if no wavelet arrives at a particular time t, then $\epsilon_t = 0$.

In our discussion of the nature of the seismic trace, we shall call the knife-sharp impulses ϵ_t "random variables." Our use of the term "random variable ϵ_t" does not imply that the variable ϵ_t is one whose value is uncertain and can be determined by a "chance" experiment. That is, the variable ϵ_t is not random in the sense of the frequency interpretation of probability (Cramer, 1946), but is fixed by the geologic structure. Frechet (1937) describes this type of variable as "nombre certain" and "function certaine" and Neyman (1941) translates these terms by "sure number" and "sure function." Another example of a "sure number" is the ten millionth digit of the expansion $e = 2.71828\cdots$, which, although unknown, is a definite fixed number. Since the knowledge of working geophysicist about the entire deterministic setting is far from complete, we shall treat this incomplete knowledge from a statistical point of view. We thus call ϵ_t a "random variable," although we keep in mind that it is a "sure number." Further discussions about this general type of problem may be found in the statistical literature with discussions about the theorem of the English clergyman Thomas Bayes and with discussions about statistical estimation (Cramer, 1946; Jeffreys, 1939). The relationship of the use of Bayes' Theorem in statistical estimation to other methods of statistical estimation is discussed by the author in his SB thesis (MIT, 1950).

Without loss of generality, we may center the knife-sharp impulses ϵ_t so that their mean $\mathbb{E}(\epsilon_t)$ is equal to zero. Nevertheless, the following discussions may be carried out, by some minor modifications, without centering the ϵ_t.

Our assumption about the unpredictability of the arrival times and strengths of wavelets means mathematically that the knife-sharp impulses ϵ_t are mutually uncorrelated random variables, that is

$$\mathbb{E}(\epsilon_t \epsilon_s) = 0 \quad \text{for } s \neq t \quad (6.21)$$

An explanation of the expectation symbol $\mathbb{E}$ is given in Section 4.2; of mutually uncorrelated variables, Section 4.5, of this thesis (MIT 1954).. Our assumption that the impulses ϵ_t are mutually uncorrelated with each other is an orthogonality assumption, and is a weaker assumption than the assumption that the ϵ_t are statistically independent, which we need not make.

Returning again, for the moment, to our discussion about the "sure" nature of the knife-sharp impulses ϵ_t, we see that the assumption that they are mutually uncorrelated in time and in strength does not hold in a completely deterministic system. Nevertheless, such an assumption is a reasonable one again for the working geophysicist whose knowledge of the entire deterministic setting is far from complete, and who is faced with essentially a statistical problem.

In other words, we assume that knowledge of the arrival time and strength of one wavelet does not allow us to predict the arrival time and strength of any other wavelets. In particular, we assume that an arrival time and magnitude of a reflection from a certain reflecting horizon does not allow us to predict the arrival time and magnitude of a reflection from a deeper reflecting horizon.

The use of AVC recordings means mathematically that the strengths ϵ_t have a constant variance $\mathbb{E}(\epsilon_t^2)$, in which case the power spectrum is white; that is, $\Phi_\epsilon(\omega) =$ constant (as described in Section 4.5 of this thesis). Without loss of generality we shall take the variance to be unity,

$$\mathbb{E}(\epsilon_t^2) = 1 \quad (6.211)$$

in which case $\Phi_\epsilon(\omega) = 1$.

Finally, since we assume that the seismogram trace x_t is additively composed of wavelets, all with shape b_t, and strengths ϵ_t, we may write this wavelet complex mathematically as

$$x_t = \sum_{s=0}^{\infty} b_s \epsilon_{t-s} \quad \text{for } t_1 \leq t \leq t_2 \tag{6.22}$$

where the time interval $t_1 \leq t \leq t_2$ comprises our basic section of seismic trace. This equation includes tails of wavelets with shape b_t, these wavelets being due to impulses

$$\epsilon_{t_1-1}, \epsilon_{t_1-2}, \cdots$$

which occur before time t_1. Equation (6.22) is illustrated in Figure 7, in which the top diagram shows the knife-sharp impulses ϵ_t, the center diagram shows the wavelets b_t weighted by these impulses, and the bottom diagram shows the seismic trace x_t which is obtained by adding the wavelets of the center diagram.

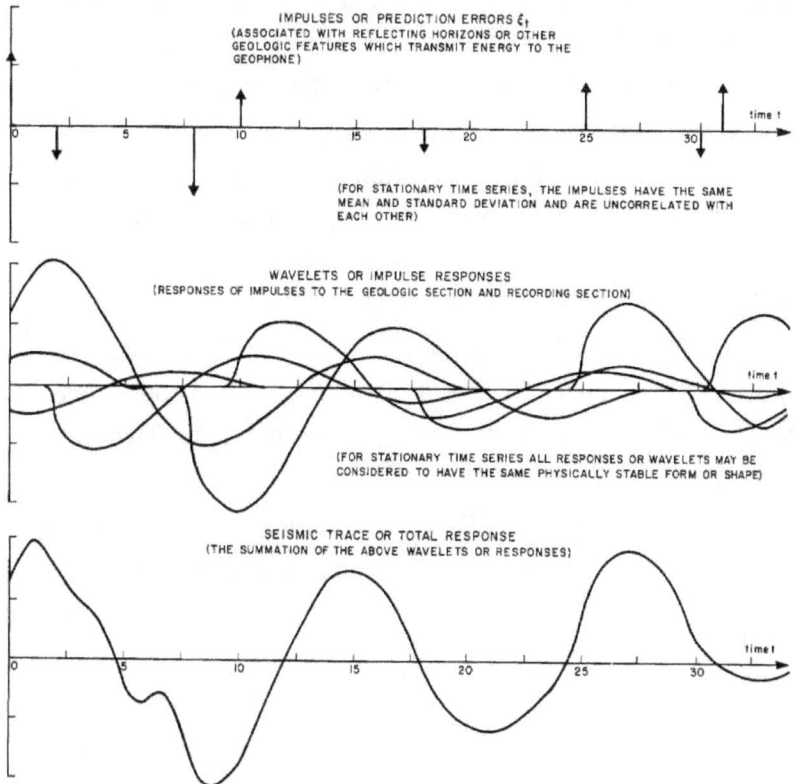

Figure 7. The predictive decomposition of a seismic trace

For the purposes of our theoretical discussion, let us assume that our assumptions about the time series x_t, equation (6.22), now hold for all time. That is, we consider the mathematical abstraction in which equation (6.22) holds for all t, where ϵ_t now represents a stationary mutually uncorrelated process (Section 4.5 of this thesis). Thus equation (6.22) becomes

$$x_t = \sum_{s=0}^{M} b_s \epsilon_{t-s} \quad \text{for } -\infty \leq t \leq \infty \quad (6.221)$$

Equation (6.221) is the mathematical representation of the Predictive Decomposition Theorem of Wold (1938). It is a convolutional model of a stationary time series x_t with an absolutely continuous spectral distribution function. For further discussion of this result, see Section 5.2 of this thesis. Thus the infinite time series x_t given by equation (6.221) is a stationary time series with an absolutely continuous spectral distribution. The finite time series x_t given by equation (6.22) represents a finite section of the infinite time series (6.221).

In equation (6.221), the minimum-phase wavelet b_t represents the "dynamics" of the seismic trace x_t, whereas the impulses ϵ_t represent the "random" nature of the seismic trace x_t. The basic problem which we wish to consider consists of the separation of the dynamic from the random components of the seismic trace.

Will the computation of the Fourier transform of the trace effect this separation? The answer is no, because that merely transforms time information into equivalent frequency information. As an illustration, let us consider the following example.

To avoid difficulties with end effects, let us assume, for this example, that the wavelet b_t damps sufficiently rapidly so that we may let

$$b_t = 0 \quad \text{for } t > M \quad (6.23)$$

Then the convolutional model becomes

$$x_t = \sum_{s=0}^{M} b_s \epsilon_{t-s} \quad (6.231)$$

Also, for this example, let us assume the trace for $t_1 \leq t \leq t_2$ consists of only those responses to impulses ϵ_t which arrive for times

$$t_1 \leq t \leq t_2 - M$$

The Fourier transform of this section of the trace becomes

$$X(\omega) = \sum_{t=t_1}^{t_2} x_t\, e^{-i\omega t} = \sum_{t=t_1}^{t_2}\sum_{s=0}^{M} b_s\, \epsilon_{t-s}\, e^{-i\omega t}$$

$$X(\omega) = \sum_{s=0}^{M} b_s\, e^{-i\omega s} \sum_{t=t_1}^{t_2} \epsilon_{t-s}\, e^{-i\omega(t-s)}$$

$$X(\omega) = \left(\sum_{s=0}^{M} b_s\, e^{-i\omega s}\right)\left(\sum_{t=t_1}^{t_2-M} \epsilon_t\, e^{-i\omega t}\right) = B(\omega)E(\omega) \quad (6.232)$$

where $B(\omega)$ is the Fourier transform, or spectrum, of the wavelet and $E(\omega)$ is the Fourier transform of a realization of the uncorrelated knife-sharp impulses. Although the Fourier transform $X(\omega)$ contains the dynamic and random elements of a seismic trace, it does not help us to separate the dynamic component $B(\omega)$ from the random component $E(\omega)$ since $X(\omega)$ is the product of the two.

In order to separate the random components ϵ_t from the dynamic component b_t of the seismic trace one may use the statistical method of averaging. The basic deconvolution approach from a theoretical point of view consists of the following operations on the mathematical abstraction of the seismic trace (i.e., the stationary time series x_t, given by equation (6.221)):

(1) Average out the random components ϵ_t so as to yield the wavelet shape b_t.

(2) Using the wavelet shape thus found, remove this wavelet shape from the trace, thereby leaving, as a residual, the random components ϵ_t (which are the prediction errors for prediction distance $\alpha = 1$.)

If one wishes to filter the seismic trace (see Section 5.4 of this thesis) one further step is added, namely:

(3) Filter the prediction errors ϵ_t by means of a linear operator q_t so as to approximate the desired output or message $m_{t+\alpha}$. That is, compute

$$\widehat{m}_{t+\alpha} = \sum_{n=0}^{\infty} q_{n+\alpha}\, \epsilon_{t-n} \quad (5.441)$$

which is the optimum filtered time series in the least-squares sense. In Section 5.4 of this thesis we describe how the linear operator q_t is determined from the spectra and cross spectra of message and noise.

The theoretical procedure for carrying out these operations has been treated in detail in our discussion of stationary time series (with absolutely continuous spectral distributions) in the preceding chapters of this thesis. There it is shown that the power spectrum $\Phi(\omega)$ of the time series x_t is equal to the product of the energy spectrum $|B(\omega)|^2$ of the wavelet b_t multiplied by the power spectrum $\Phi_\epsilon(\omega)$ of the random components ϵ_t. According to equation (6.211), $\Phi_\epsilon(\omega) = 1$. Thus we have

$$\Phi(\omega) = |B(\omega)|^2 \Phi_\epsilon(\omega) = |B(\omega)|^2$$

Let us review this theoretical procedure for infinite stationary time series.

(1) Compute the autocorrelation function of the time series

$$\phi(\tau) = \mathbb{E}(x_t\, x_{t+\tau}) = \lim_{T\to\infty} \frac{1}{2T+1} \sum_{t=-T}^{T} x_t\, x_{t+\tau}$$

$$= \sum_{t=0}^{\infty} b_t\, b_{t+\tau} = b_0\, b_\tau + b_1\, b_{\tau+1} + b_2\, b_{\tau+2} + \qquad (5.232)$$

This computation averages out the random elements ϵ_t and preserves the dynamic elements b_t in the form of the autocorrelation function

$$\phi(\tau) = \sum_{t=0}^{\infty} b_t\, b_{t+\tau}$$

of the wavelet. That is, the autocorrelation of the time series x_t is the same function as the autocorrelation function of the wavelet b_t.

From this autocorrelation function, compute the shape b_t of the wavelet in the following manner. Take the Fourier transform of the autocorrelation function to get the power spectrum $\Phi(\omega)$ of the time series x_t. The power spectrum $\Phi(\omega)$ is equal to the energy spectrum $|B(\omega)|^2$ of the wavelet b_t; that is

$$\sum_{n=-\infty}^{\infty} \phi(\tau) e^{-i\omega\tau} = \Phi(\omega) = |B(\omega)|^2 \qquad (6.24)$$

where

$$B(\omega) = \sum_{t=0}^{\infty} b_t e^{-i\omega t} \qquad (6.241)$$

is the Fourier transform of the wavelet b_t.

The Fourier transform (6.241) may be written as

$$B(\omega) = |B(\omega)| e^{i\theta(\omega)}$$

From equation (6.24) we know that

$$|B(\omega)| = \sqrt{\Phi(\omega)}$$

We have determined $|B(\omega)|$ but not $B(\omega)$. Although there are many different wavelet shapes that have the same energy spectrum $|B(\omega)|^2$, only one of these wavelet shapes is minimum-phase. Therefore it is not unreasonable to assume that this unique minimum phase wavelet is the wavelet generated by a physical phenomenon and the others are not.

The Fourier transform $B(\omega)$ of this minimum-phase wavelet may then be found by the Fejer or Kolmogorov method of factoring the power spectrum (see Section 5.1 of this thesis) expressed by

$$\Phi(\omega) = |B(\omega)|^2 = B(\omega) \overline{B(\omega)} \qquad (5.181)$$

where $B(\lambda)$ is required to have no singularities or zeros in the lower half λ plane, where $\lambda = \omega + i\sigma$. In the language of the engineer, $B(\omega)$ is a transfer function with minimum phase-shift characteristic $\theta(\omega)$ given by

$$\theta(\omega) = -\frac{1}{\pi} \sum_{t=1}^{\infty} \sin \omega t \int_0^\pi \cos ut \, \log \Phi(u) \, du$$

Using this $\theta(\omega)$ together with $\sqrt{\Phi(\omega)}$, we can obtain the desired

$$B(\omega) = \sqrt{\Phi(\omega)} \, e^{i\theta(\omega)}$$

Having thus determined $B(\omega)$, the minimum-phase wavelet b_t is given by

MIT AND THE BIRTH OF DIGITAL SIGNAL PROCESSING

$$b_t = \frac{1}{2\pi} \int_{-\pi}^{\pi} B(\omega) e^{-i\omega t} d\omega \qquad (6.242)$$

(2) From this wavelet shape b_t we find the inverse wavelet shape a_t where a_t is equal to zero for t less than zero. If we let the b_t represent the coefficients of a linear operator, then the a_t are the coefficients of the inverse linear operator (Section 2.7 of this thesis). Thus the values of a_t are found by

$$a_t = 0 \quad \text{for } t < 0, \qquad a_0 b_0 = 1$$

$$\sum_{s=0}^{t} a_s b_{t-s} = 0 \quad \text{for} \quad t = 1, 2, 3, \cdots \qquad (2.785)$$

Since the wavelet b_t is minimum-phase, the inverse wavelet a_t is also minimum-phase. Let $A(\omega)$ be the Fourier transform of a_t; that is,

$$A(\omega) = \sum_{s=0}^{\infty} b_s e^{-i\omega s} \qquad (2.791)$$

Then $A(\omega)$ and $B(\omega)$ are related by

$$A(\omega) = \frac{1}{B(\omega)} \qquad (2.795)$$

and $A(\omega)$ also has minimum phase-shift characteristic. The reciprocal of $|A(\omega)|$ then gives $|B(\omega)|$ which is the absolute value of the wavelet spectrum.

We use the inverse wavelet shape a_t to remove the wavelets, which are of shape b_t, from the time series x_t, by compressing the wavelets into the knife-sharp impulses ϵ_t. That is, the linear operator a_t is the prediction error operator for unit prediction distance, and the prediction errors ϵ_t are yielded by the deconvolution computation

$$\sum_{s=0}^{\infty} a_s x_{t-s} \qquad (6.25)$$

To see that this computation does yield the ϵ_t, we use the convolutional model (6.221) for x_t and thus obtain

$$\sum_{s=0}^{\infty} a_s x_{t-s} = \sum_{s=0}^{\infty} a_s \sum_{\tau=0}^{\infty} b_\tau \epsilon_{t-s-\tau}$$

$$= \sum_{s=0}^{\infty} a_s \left[\sum_{n=s}^{\infty} b_{n-s} \epsilon_{t-n} \right] \qquad (6.251)$$

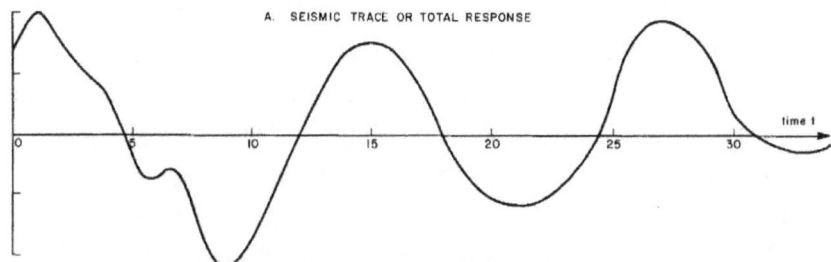

A. SEISMIC TRACE OR TOTAL RESPONSE

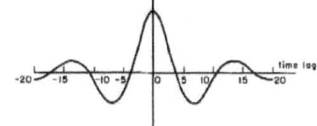

B. AUTOCORRELATION FUNCTION OF TRACE
(UNIQUELY DETERMINED FROM INFINITE TIME SERIES, BUT MAY BE ESTIMATED FROM FINITE SECTION OF TRACE)

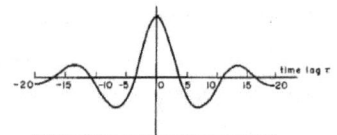

C. AUTOCORRELATION FUNCTION OF INDIVIDUAL WAVELET
(SAME AS AUTOCORRELATION FUNCTION OF TRACE)

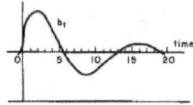

D. PHYSICALLY STABLE FORM OR SHAPE OF INDIVIDUAL WAVELET
(UNIQUELY DETERMINED FROM WAVELET AUTOCORRELATION BY METHOD OF THE FACTORIZATION OF THE SPECTRUM)

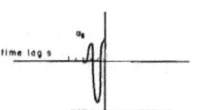

E. STABLE PREDICTION OPERATOR WITH MINIMUM PHASE CHARACTERISTIC
(UNIQUELY DETERMINED FROM WAVELET FORM)

F. IMPULSE OR PREDICTION ERROR
YIELDED BY PREDICTION OPERATOR ACTING ON WAVELET. THIS IMPULSE OCCURS AT THE ARRIVAL TIME OF WAVELET

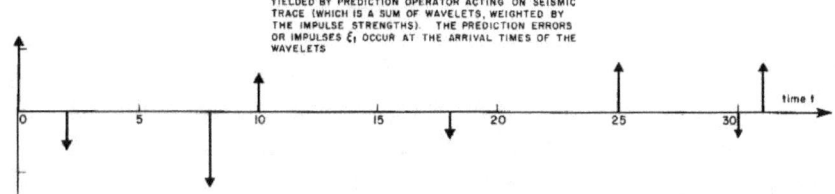

G. SERIES OF IMPULSES OR PREDICTION ERRORS YIELDED BY PREDICTION OPERATOR ACTING ON SEISMIC TRACE (WHICH IS A SUM OF WAVELETS, WEIGHTED BY THE IMPULSE STRENGTHS). THE PREDICTION ERRORS OR IMPULSES ξ_t OCCUR AT THE ARRIVAL TIMES OF THE WAVELETS

Figure 8. Analysis of a seismic trace by a prediction operator

Recalling that $b_t = 0$ for $t < 0$, and using equation (2.785), we have

$$\sum_{s=0}^{\infty} a_s x_{t-s} = \sum_{n=0}^{\infty} \left[\sum_{s=0}^{\infty} a_s b_{n-s} \right] \epsilon_{t-n} = \epsilon_t \qquad (6.252)$$

Thus we have

$$\sum_{s=0}^{\infty} a_s x_{t-s} = \epsilon_t \qquad (6.253)$$

which is the prediction error, or knife-sharp impulse.

Thus by these theoretical steps we may separate the dynamic component, represented by the response function or wavelet shape b_t from the random component, represented by the knife-sharp impulses ϵ_t which represent arrival times and strengths of the wavelets which comprise the time series. These theoretical steps are illustrated in Figure 8. In this figure, as in others, we plot the discrete time functions as points and then draw smooth curves through these points.

The practical solution of the problem of separating the dynamic and random components of a finite section of seismic trace involves statistical estimation. One method consists of estimating the prediction error operator, or inverse wavelet shape, directly from the finite section of seismic trace. For this purpose one may use the Gauss method of least squares as described in Wadsworth, Robinson, Bryan, and Hurley (1953). Since the method described there is more general, let us write down the equations to be used for our specific problem in which we consider only one trace x_t for a prediction distance equal to one.

Then utilizing our notation, equation (37) of Wadsworth, Robinson, Bryan, and Hurley (1953) becomes for the special case of our problem:

$$\hat{x}_{t+1} = c + \sum_{s=0}^{M} k_s x_{t-s} \qquad (6.26)$$

According to our convention, we have let the digitization spacing $h = 1$ time unit (where the time unit is 2.5 ms) so that the running index t is an integer. The constant c appears in equation (6.26) to take account of the mean value of the time series, since we do not require the mean to be zero.

The operator time interval [also known as the time gate] is chosen to be the time interval of the section of our hypothetical trace which we assume to be a section of a stationary time series. The operator time interval consists of those n values of x_t from $t = N$ to $t = N + n - 1$. The normal equations are

$$cn + \sum_s k_s \sum_t x_{t-s} = \sum_t x_{t+1}$$

$$c \sum_t x_{t-r} + \sum_s k_s \sum_t x_{t-r} x_{t-s} = \sum_t x_{t-r} x_{t+1}$$

$$\text{for } r = 0, 1, \cdots, M \qquad (6.261)$$

The summation is taken over the operator time interval; that is, the summation index t runs from $t = N - 1$ to $t = N + n - 2$ since the prediction distance $\alpha = 1$. All summations on the index s are for $s = 0$ to $s = M$. The solution of the normal equations (6.261) yields the operator coefficients

$$c, k_0, k_1, \cdots, k_M$$

Then the inverse wavelet shape a_t is given by equation (2.281) which is

$$a_0 = 1, a_1 = -k_0, a_2 = -k_1, \cdots, a_m = -k_M \qquad (2.281)$$

with $m = M + 1$. As we have noted in Section 2.2 of this thesis, although both a_t and k_t represent the coefficients of the same operator but in different form, as seen by equation (2.281), we call a_t the standard form of the prediction error operator. The constant c of equation (6.26), which adjusts for the mean value of the empirical trace, is not used in determining- the shape a_t of the wavelet. Since $a_0 = 1$, the inverse wavelet a_t may be called a "unit" inverse wavelet. The convolution of the inverse operator a_t [also known as the deconvolution operator] with the seismic trace x_t yields the prediction error trace ϵ_t [also known as the deconvolved trace], as seen from equation (6.252).

The shape b_t of the wavelet may be readily computed by means of equations (2.785) of this thesis, which are

$$a_0 b_0 = 1, \quad \sum_{s=0}^{t} a_s b_{t-s} = 0 \text{ for } t = 1, 2, 3, \cdots \qquad (2.785)$$

Since we let $a_0 = 1$, we have $b_0 = 1$. Thus given the set a_s, the set b_s may be uniquely determined, and vice versa. Equation (2.785) may be rewritten in terms of the prediction operator k_s of equation (6.26) as

$$b_t = 0 \text{ for } t < 0, \quad b_0 = 1,$$

$$b_{t+1} = -\sum_{s=1}^{m} a_s b_{t+1-s} = \sum_{s=0}^{M} k_s b_{t-s} \text{ for } t > 0 \quad (6.262)$$

That is, the wavelet shape b_t for $t > 0$ is determined by successive step-by-step predictions from its past values, where we let the initial values be $b_t = 0$ for < 0 and $b_0 = 1$.

As we have seen, the Gauss method of least squares described yields an empirical estimate of the theoretical prediction error operator, or inverse wavelet shape. This empirical estimate has certain optimum statistical properties under general conditions. For a treatment of the optimum properties of linear least-squares estimates, see Wold (1953).

A good estimate of the prediction operator should yield prediction errors which are not significantly autocorrelated. In other words, the prediction errors ϵ_t should be mutually uncorrelated at some pre-assigned level of significance. Let it be noted that we are confining our attention to the hypothetical section of the trace which we assumed to be a section of a stationary time series; that is, we are dealing with the prediction errors in the so-called operator time interval. For example, if the prediction errors are significantly autocorrelated, more coefficients may be required in the empirical prediction operator.

In Figure 9, in the left hand diagram, we show the prediction error (i.e., deconvolution) operator a_s computed for trace N650 for the time interval 0.350 sec to 0.475 sec on MIT Record No. 1 (supplied by the Magnolia Petroleum Co.). This seismogram is illustrated and described in Wadsworth, Robinson, Bryan, and Hurley (1953). In the computation of this inverse wavelet shape a_s, we used equations (6.261) to find the k_s, and then used equations (2.281) to find the a_s. In the right hand diagram of Figure 9, we show the inverse prediction error operator, which is the shape of the wavelet b_t. The shape of the wavelet was

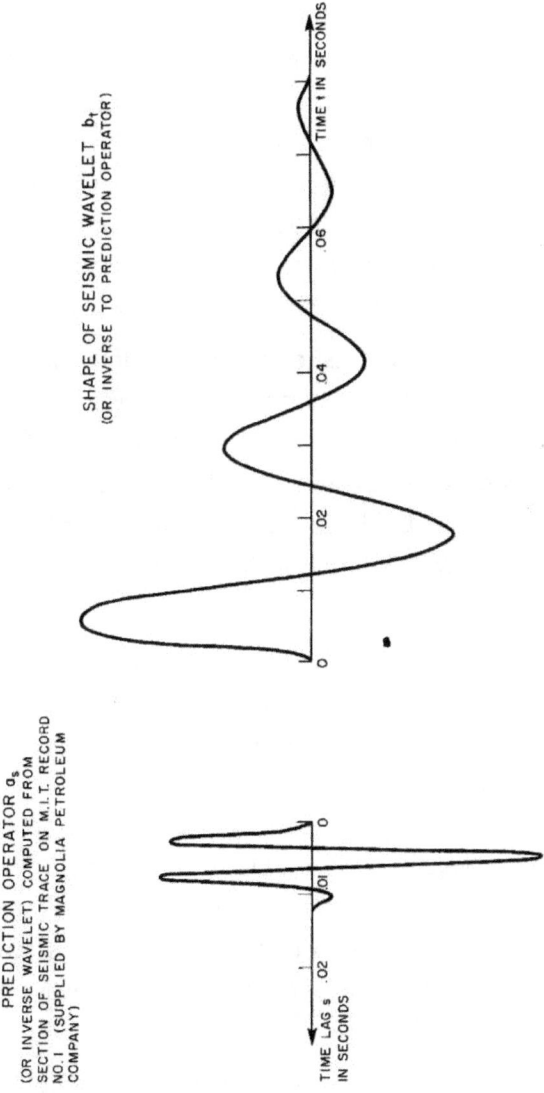

Figure 9. Computation of a wavelet from a section of trace on MIT Record No. 1. [In today's language, PREDICTION OPERATOR should read PREDICTION ERROR OPERATOR. The wavelet is Irene's minimum-phase wavelet.]

"predicted" by means of equation (6.262). In these computations, we used discrete time series with the spacing increment $h = 2.5$ ms. In plotting a_s and b_t we followed our usual procedure which is to plot discrete time functions, such as a_s and b_t as discrete points, and then to draw a smooth curve through these discrete points. Also in Figure 9, the time axes are shifted by one discrete time unit (which is 2.5 ms), which is not the convention we have used in our other figures. Thus in Figure 9, $a_{-1} = 0$ is plotted at time lag $s = 0$, and $a_0 = 1$ is plotted at time lag $s = 0.0025$ sec. Similarly $b_{-1} = 0$ is plotted at time $t = 0$ sec and $b_0 = 1$ is plotted at time $t = 0.0025$ sec. As is our usual convention, the prediction error operator a_s is plotted in the reverse manner; that is, the time lag s runs in the positive direction toward the left.

Here we have described a statistical method to determine the shape of a seismic wavelet. Alternatively, from other considerations, one may know the shape of the seismic wavelet. Then the prediction error operator, or inverse wavelet shape, may be computed by means of equations (2.785). [The prediction error operator a_s is the *deconvolution operator,* and the prediction error trace ϵ_t is the *deconvolved trace.*]

So far we have confined ourselves to a section of trace which we assume to be approximately stationary. The prediction error operator transforms this section of trace into the uncorrelated prediction errors ϵ_t, the mean square value of which is a minimum. As we have seen, from past values of the trace, the operator cannot predict the initial arrival of a new wavelet, and thus a prediction error ϵ_t is introduced at the arrival time of each wavelet. Nevertheless, for times subsequent to the arrival time of the wavelet, the prediction error operator, which is the inverse to the wavelet, can perfectly predict this wavelet, thereby yielding zero error of prediction.

Nevertheless, a seismic trace as recorded is not made up of wavelets which have exactly the same form and which differ only in amplitudes and arrival times. Thus if a prediction error operator, which is the unique inverse of a certain wavelet shape, encounters a different wavelet shape, the prediction error will no longer be an impulse, but instead will be a transient time function. Thus the prediction errors yielded by this prediction error operator acting on a time series additively composed of wavelets of different shapes will not have a minimum mean square value. Since reflected wavelets in many cases have different shapes than the wavelets comprising the seismic trace in

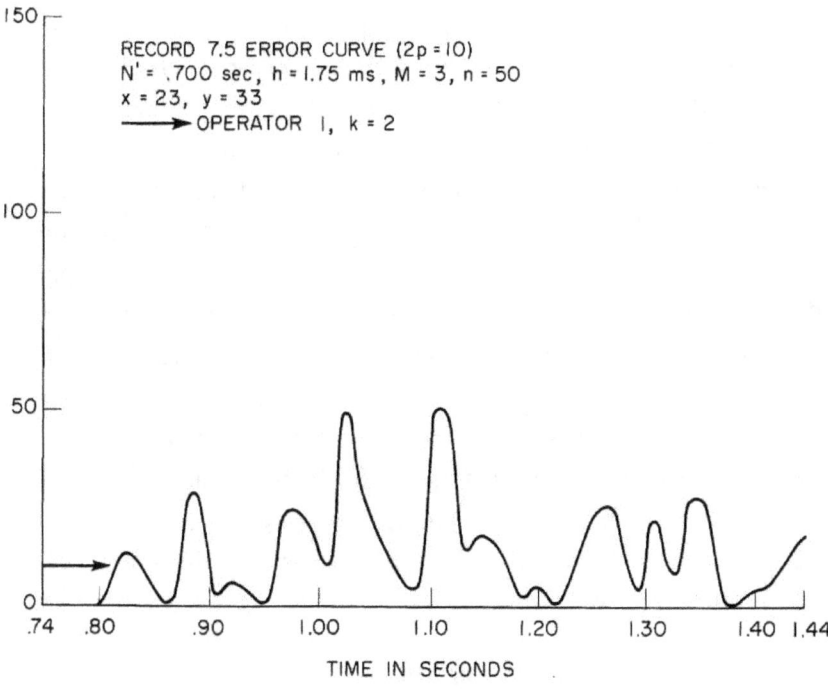

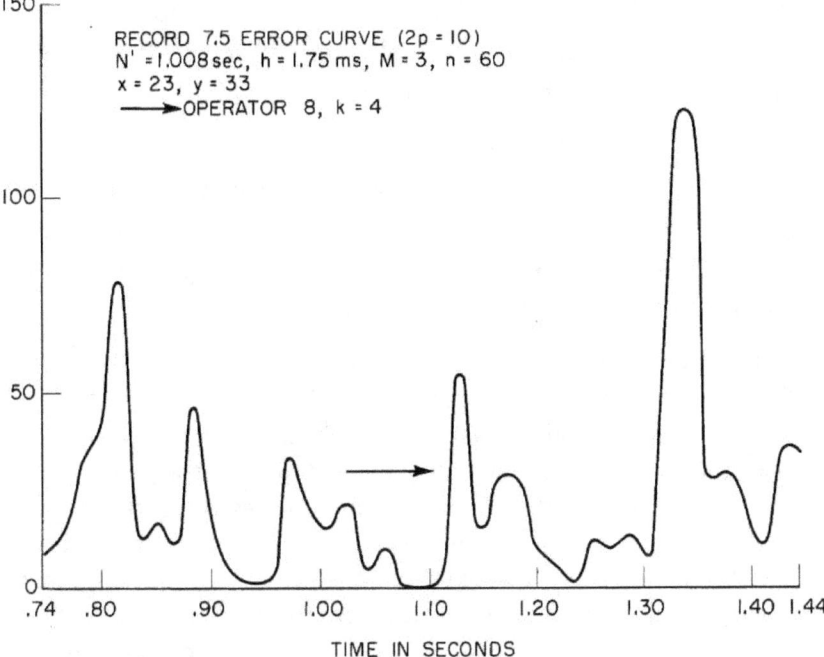

Figures 10 and 11. Prediction error curves. The peaks indicate the reflections.

a given non-reflection interval, a prediction error operator determined from this non-reflection interval will yield high errors of prediction at such reflections. Such a procedure provides a method for the detection of reflections (Wadsworth, Robinson, Bryan, and Hurley, 1953). In Figures 10 and 11, running averages of the squared prediction errors are plotted. The peaks on these prediction error curves indicate reflections on the seismogram. Since two-trace operators were used, the empirical coherency existing between the two traces was utilized in the determination of these prediction errors. The arrow indicates the operator time interval. Further description of these figures is given in MIT GAG Report No. 6. Since only the information existing in the operator time interval is used in the determination of linear operators by this method, one may expect greater resolving power if more information on the seismogram is utilized in the determination of various other types of operators.

Mike Perz and Gary Margrave

On May 7, 1955 I submitted to *Geophysics* the above Chapter VI together with the necessary mathematics from Chapters I through V. The paper was deemed too long to publish so I cut out the mathematics. Because of backlog, it was finally published in 1957.

Mike Perz (ARCIS Seismic Solutions) and Gary Margrave (University of Calgary) wrote "Assessing the Impact of Robinson's First Deconvolution Paper" (*CSEG Recorder*, February 2005). Some of the highlights of this beautifully written and instructive paper are:

> It is difficult to overestimate the importance and depth of Enders Robinson's contributions to geophysical signal processing. Early among these, and indispensable in modern data processing, is the theory of spiking deconvolution as set forth in a classic paper in 1957 titled *Predictive Decomposition of Seismic Traces*. Appearing in *Geophysics* and authored solely by Robinson, this paper triggered a revolution in data processing by demonstrating that (1) the seismic wavelet could be estimated directly from the seismic trace itself and (2) an effective inverse for this wavelet can be designed and applied. This process has come to be called deconvolution and the resulting deconvolved trace usually shows a dramatic improvement in

temporal resolution as well as an apparent reduction in strength of coherent noises like ground roll."

Today, the details of deconvolution theory are a standard component in the education of a geophysicist and it is not our intent to review them here. Rather, we intend to concentrate on Robinson's paper and, with the acute perspective of hindsight, comment upon the clarity of its vision and its relevance to our present understanding. In doing so, we hope to answer such questions as (1) Who were Robinson's major influences in developing his 1957 paper? (2) Did Robinson get it right from the beginning or were there significant missing ingredients? and (3) Do we understand the theory today in the same terms as were originally used?

In a section entitled Wavelet Theory, Robinson describes the 1940 theory of Ricker that the seismogram can be regarded as an elaborate wavelet complex. Though never stated explicitly, this theory is clearly a convolutional theory as it explicitly builds a seismogram as a superposition of many wavelets, with different strengths and arrival times. It is also recognizably a nonstationary convolutional theory since it recognizes the wavelet as a traveling wavelet, the shape of which is determined by the nature of the absorption spectrum of the earth for elastic waves. Robinson goes on to explicitly invoke the minimum-phase assumption, much as we would today, and characterizes such a wavelet as a one-sided transient which damps with a certain degree of rapidity. Later, the minimum-phase wavelet is defined as that which results from the Wold-Kolomogorov factorization of the power spectrum. Robinson's equation for the minimum phase $\theta(\omega)$ is an explicit prescription for constructing the minimum-phase spectrum and it is completely equivalent to our modern statement that the minimum-phase spectrum is the Hilbert transform of the logarithm of the amplitude spectrum. However, Robinson's equation for the minimum phase $\theta(\omega)$ is presented in a form that would be very efficient for numerical calculations in that it takes advantages of available symmetries such as the power spectrum being the same for positive and negative frequencies.

Interestingly, it appears that Robinson's primary motivation for invoking the minimum phase assumption was not physical reasonableness, but rather the fact that minimum phase sequences possessed certain desirable mathematical properties which we'll describe below. In 1966, over ten years after completing his thesis, he provided a physical justification for the minimum phase assumption by noting that certain physically plausible wave propagation effects—notably the source ghost and the water-bottom reverberation pulse-train, could be reasonably modeled via minimum phase sequences. In that same *Geophysics* paper he writes, "At the time of the writing of my thesis at MIT, some (physically-rooted) checks of the above (minimum phase) model were made but were necessarily of a limited nature because of the data and computing facilities..."

And so we see that by connecting the seismic trace to Wold via Ricker, Robinson has made the two key assumptions which together form the cornerstone of today's deconvolution algorithms, namely that the wavelet is minimum phase, and that the reflectivity is white (i.e. unpredictable). Today, we're sufficiently familiar with the concept of white reflectivity that we often take the idea for granted, but one could imagine that the notion of ascribing a purely random behaviour to something as intrinsically deterministic as geology would be quite a stretch in 1957! At least it would have been without the mathematical bolstering of the Wold decomposition theorem.

After this analysis, we feel that the deconvolution algorithm of Robinson's 1957 paper has withstood the "test of time" far better that the vast majority of geophysical innovations. It is not just that the physical analysis was sound, but also the proposed algorithm has proven very robust in the presence of the many complications found in real data. While our modern understanding has matured somewhat and both surface-consistent and nonstationary extensions of Robinson's method have been developed, we are still fundamentally in agreement with the trace model (proposed by Ricker and championed by Robinson) and with the perspective of minimum-phase prediction error filtering for wavelet deconvolution. It is certainly true that these ideas are approximations to the complexity of real seismic data, and approximations, by their very nature, are not at all unique. However, in a field changing as rapidly as exploration geophysics, the longevity of an idea is often directly related to the intangible art and wisdom of the approximation. By thriving for 50 years, Robinson's deconvolution demonstrates a very high level of both wisdom and art.

Chapter 7. Andre Kolmogorov

Andre Kolmogorov (1903-1987), one of the foremost mathematicians of the twentieth century and of all time, made significant advances in the fields of probability theory, topology, intuitionistic logic, turbulence, classical mechanics, algorithmic information theory and computational complexity.

At the age of 18, he was internationally recognized for the construction of a Fourier series that diverges almost everywhere. At the age of 22, he published his famous work in intuitionistic logic, *On the principle of the excluded middle*. In 1931, he published *About the Analytical Methods of Probability Theory*. In 1933, Kolmogorov published his book, *Foundations of the Theory of Probability*, which laid the modern axiomatic foundations of probability theory. It established his reputation as the world's leading expert in this field.

In 1936 Kolmogorov generalized the Lotka–Volterra model of predator-prey systems. In a 1938 paper, Kolmogorov established the basic theorems for smoothing and predicting stationary stochastic processes—a paper that preceded the work of Norbert Wiener in that area.

In his study of stochastic processes (random processes), especially Markov processes, Kolmogorov and the British mathematician Sydney Chapman independently developed the pivotal set of equations in the field, which have been given the name of the Chapman–Kolmogorov equations.

Kolmogorov then focused his research on turbulence, where his publications significantly influenced the field. In classical mechanics, he is best known for the Kolmogorov–Arnold–Moser theorem, first presented in 1954 at the International Congress of Mathematicians. In 1957, working jointly with his student, V. I. Arnold, he solved a particular interpretation of Hilbert's thirteenth problem. Around this time he also began to develop, and was considered a founder of, algorithmic complexity theory - often referred to as Kolmogorov complexity theory. In his later years, he devoted much of his effort to the mathematical and philosophical relationship between probability theory in abstract and applied areas

From Russia, Professor Kolmogorov visited Sweden in 1961. Professor Wold introduced him to me. We discussed my work on digital filtering and deconvolution, and he then asked me to publish my findings in the Russian journal *Theory of Probability and its Applications* which he founded. I prepared a manuscript and Prof. Wold sent it to him. In answer, he wrote this letter.

> Dear Prof. Wold,
>
> Thank you very much for your nice letter of 26 April 1961 and the enclosure. The paper of Prof. Robinson sent by you was discussed in the editorial office of the journal. For specialists in the field of stationary processes the statement of the problem contained in Theorem 12 is new because the quantity $\hat{x}_{t+a}$ of the best linear prediction of any regular process x_t is compared in a unique way with the series $\sum k_s x_{t-s}$ such that the difference
>
> $$x_{t+a} - \sum k_s x_{t-s}$$
>
> belongs to the linear completion $\Theta(t - N + 1)$ of the values $x_k, -\infty < k < t - N$ where by virtue of the regularity of x_t
>
> $$\bigcap_{N \geq 0} \Theta(t - N + 1) = 0$$

Chapter 7. Andre Kolmogorov

I request that you convey this information to Prof. Robinson.

Sincerely yours

A. Kolmogorov

The result was that Professor Kolmogorov published my paper *Properties of the Wold decomposition of stationary stochastic processes*. This paper cleared up the mathematical loose ends in deconvolution once and for all. Wold, Wiener and Kolmogorov all concurred on this decision. The title page of the article is

ТЕОРИЯ ВЕРОЯТНОСТЕЙ И ЕЕ ПРИМЕНЕНИЯ

Том VIII

2

НЕКОТОРЫЕ СВОЙСТВА РАЗЛОЖЕНИЯ ВОЛЬДА СТАЦИОНАРНЫХ СЛУЧАЙНЫХ ПРОЦЕССОВ

Е. А. РОБИНСОН (УПСАЛА, ШВЕЦИЯ)

(*Резюме*)

Основные результаты работы (теоремы 11—13) касаются вопроса о представимости величин $\hat{x}_{t+a}$ (являющихся наилучшим прогнозом значений x_{t+a} стационарного в широком смысле процесса по значениям x_s, $s \leqslant t$) в виде ряда

$$\hat{x}_{t+a} \sim \sum_{s=0}^{\infty} k_s x_{t-s},$$

где коэффициенты k_s удовлетворяют условию: $\sum |k_s|^2 < \infty$. Предварительно приводятся некоторые свойства последовательностей $\{w_t\}$, $\sum |w_t|^2 < \infty$.

The paper reads as follows.

The wavelet

A sequence w_t of complex numbers (where t is an integer) is called a wavelet if $w_t = 0$ for $t < 0$ and $\sum_{t=0}^{\infty} |w_t|^2$. If we define the inner product between two wavelets w_t and v_t by

$$(w_t, v_t) = \sum_{t=0}^{\infty} w_t \overline{v_t}$$

then the space of all wavelets is a Hilbert space, which we shall denote by $L^2(0, \infty)$. The z-transform of a wavelet w_t is defined to be

$$W(z) = \sum_{t=0}^{\infty} w_t z^t$$

$W(z)$ is an analytic function for $|z| < 1$, and has the limit value

$$W(e^{-2\pi i f}) = \sum_{t=0}^{\infty} w_t e^{-2\pi i f t}, \text{a. e.}$$

on the unit circle $z = e^{-2\pi i f}$.

Let us consider all complex-valued functions $H(v)$ on the interval $-0.5 \leq f \leq 0.5$ such that

$$\int_{-0.5}^{0.5} |H(f)|^2 \, df < \infty$$

If we define the inner product between two such functions $H(f)$ and $G(f)$ by

$$(H(f), G(f)) = \int_{-0.5}^{0.5} H(f) \overline{G(f)} \, df$$

then the space of all such functions is a Hilbert space, which we shall denote by $L^2(df)$.

It is well-known that if w_t is a wavelet, then there is a function $W(f)$ in space $L^2(df)$ given by

$$W(f) = \underset{N \to \infty}{\text{l. i. m.}} \sum_{t=0}^{N} w_t e^{-2\pi i f t} = W(e^{-2\pi i f}), \text{a. e.} \quad (1)$$

Moreover, if the wavelets w_t and v_t yield $W(f)$ and $V(f)$, respectively, then

$$\sum_{t=0}^{\infty} w_t \overline{v_t} = \int_{-0.5}^{0.5} W(f)\overline{V(f)}\,df$$

This equation is known as *Bessel's equality* if $w_t = v_t$ (and consequently $W(f) = V(f)$). Otherwise, it is known as *Parseval's equality*.

The function $W(f)$, given by equation (1), may be written in polar form as

$$W(f) = |W(f)|e^{i\psi(f)}$$

where $|W(f)|$ is called the *gain*, and $\psi(f)$ the *phase-shift*. The *group-delay* is defined to be

$$\tau_g = -\frac{d\psi(f)}{df}, \quad \text{a. e.}$$

The *all-pass z-transform* is defined as the function

$$\mathcal{P}(z) = cz^m \prod_k \frac{z_k - z}{1 - \overline{z_k}z} \frac{|z_k|}{z_k} \exp\left\{-\int_{-0.5}^{0.5} \frac{e^{2\pi i f} + z}{e^{2\pi i f} - z} d\beta(f)\right\}$$

where

$|z| < 1; |c| = 1;$

m is a non-negative integer;

$\{z_k\}$ is an empty or non-empty set satisfying

$$|z_k| < 1 \quad \text{and} \quad \sum_k (1 - |z_k|) < \infty$$

and $\beta(f)$ is a real, bounded, non-decreasing function whose derivative vanishes almost everywhere.

If $|\mathcal{P}(z)| = 1$ for $|z| < 1$, then $\mathcal{P}(z)$ is called trivial: otherwise $\mathcal{P}(z)$ is called nontrivial.

The *minimum-delay z-transform* with gain $M(f)$ is defined as

$$\mathcal{B}(z) = \exp\left\{-\int_{-0.5}^{0.5} \frac{e^{2\pi i f} + z}{e^{2\pi i f} - z} \log M(f)\,df\right\}$$

where

$$|z| < 1; \quad M(f) \geq 0;$$

$$\int_{-0.5}^{0.5} |M(f)|^2 df < \infty; \quad \text{and} \quad \int_{-0.5}^{0.5} \log M(f)\, df > -\infty.$$

Canonical representation

The following *canonical representation* follows from Theorems 6.2 and 11.4 in Privalov (1956, pp. 78 and 110). See also Szego (1921), Riesz (1923), Smirnov (1928), Krylov (1939), Kolmogorov (1941), Karhunen (1949), Yaglom (1955), and Grenander and Rosenblatt (1957).

Theorem 1. $W(z)$ is the z-transform of a wavelet if and only if

$$W(z) = \mathcal{P}(z)\mathcal{B}(z), \quad |z| < 1$$

where $\mathcal{P}(z)$ is an all-pass z-transform and $\mathcal{B}(z)$ is the minimum-delay z-transform with the same gain as $W(z)$. This representation of $W(z)$ is unique.

We now derive:

Theorem 2. $\mathcal{P}(z)$ is an all-pass z-transform if and only if $\mathcal{P}(z)$ is the z-transform of a wavelet with gain equal to 1, a. e.

Proof. The gain $M(f) = 1$ has the minimum-delay z-transform $\mathcal{B}(z) = \exp 0 = 1$. If $\mathcal{P}(z)$ is all-pass, then $\mathcal{P}(z) = \mathcal{P}(z) \cdot 1$ is a canonical representation. Hence $\mathcal{P}(z)$ is the z-transform of a wavelet with gain 1. Conversely, if $\mathcal{P}(z)$ is the z-transform of a wavelet with gain 1, then its canonical representation must be $\mathcal{P}(z) = \mathcal{P}(z) \cdot 1$. Hence $\mathcal{P}(z)$ is an all-pass z-transform. Q. E. D.

By this theorem, an all-pass z-transform $\mathcal{P}(z)$ is the z-transform of a wavelet p_t which we call an all-pass wavelet. The first examples of all-pass wavelets were given by Wold (1938, p. 130). An all-pass wavelet has gain $|P(f)| = 1$, a. e. This unit gain property is the reason for using the term "all-pass." The minimum-delay wavelet with gain equal to one is the unit-impulse

$$\delta_t = 1 \text{ if } t = 0 \text{ and } \delta_t = 0 \text{ if } t \neq 0$$

Chapter 7. Andre Kolmogorov

The unit-impulse has z-transform $\mathcal{P}(z) = 1$. Using this z-transform in Theorem 1, we see that $\mathcal{B}(z)$ is the z-transform of a wavelet β_t, which we call the minimum-delay wavelet with gain $M(f)$.

The canonical representation in terms of wavelets is

Theorem 3. w_t is a wavelet if and only if

$$w_t = \sum_{s=0}^{t} \beta_s \, p_{t-s}$$

where p_t is an *all-pass wavelet* and β_t is the *minimum-delay wavelet* with the same gain as w_t. This representation of w_t is unique. Theorem 3 shows that the partial sums

$$\sum_{s=0}^{N} \beta_s \, p_{t-s}$$

for fixed t converge to the value w_t. The following theorem shows that these partial sums also converge in the mean to the wavelet w_t.

Theorem 4. In Hilbert space $L^2(0, \infty)$, we have

$$w_t = \underset{N \to \infty}{\text{l.i.m.}} \sum_{s=0}^{N} \beta_s \, p_{t-s}$$

Proof. By Bessel's equality, we have

$$\sum_{t=0}^{\infty} \left| w_t - \sum_{s=0}^{N} \beta_s \, p_{t-s} \right|^2 = \int_{-0.5}^{0.5} \left| W(f) - \sum_{s=0}^{N} \beta_s \, e^{-2\pi i f s} P(f) \right|^2 df$$

Because

$$W(f) = B(f)P(f) \quad \text{and} \quad |P(f)| = 1,$$

the above is equal to

$$\int_{-0.5}^{0.5} \left| B(f) - \sum_{s=0}^{N} \beta_s \, e^{-2\pi i f s} \right|^2 df = \sum_{s=N+1}^{\infty} |\beta_s|^2$$

which can be made as small as we please by choosing N large enough. Q. E. D.

All pass z-transform

The following three theorems can be established directly from the definition of the all-pass z-transform.

Theorem 5. The group-delay τ_g of an all-pass z-transform is positive (resp. zero) for $-0.5 < f < 0.5$ if and only if the z-transform is nontrivial (resp. trivial).

Theorem 6. The modulus $|\mathcal{P}(z)|$ of an all-pass z-transform satisfies $|\mathcal{P}(z)| < 1$ (resp. $|\mathcal{P}(z)| = 1$) for $|z| < 1$ if and only if the all-pass z-transform is non-trivial (resp. trivial).

It follows from Theorem 3 that if w_t is a wavelet, and p_t is an all-pass wavelet, then

$$v_t = \sum_{s=0}^{t} w_s \, p_{t-s}$$

is also a wavelet. By a proof similar to that of Theorem 4, it may be shown that in Hilbert space $L^2(0, \infty)$,

$$v_t = \underset{N \to \infty}{\text{l.i.m.}} \sum_{s=0}^{N} w_s \, p_{t-s}$$

Because $P(f) = 1$, a. e., application of Bessel's equality gives

$$\sum_{t=0}^{\infty} |v_t|^2 = \sum_{t=0}^{\infty} |w_t|^2$$

which says that the two wavelets v_t and w_t have the same total energy. The following theorem, however, says that the all-pass wavelet p_t delays the partial energy of v_t with respect to the partial energy of w_t.

Theorem 7. Let w_t be a wavelet, and p_t be an all-pass wavelet. Then the partial energy of the wavelet

$$v_t = \sum_{s=0}^{t} w_s \, p_{t-s}$$

never exceeds the partial energy of w_t; i. e.

Chapter 7. Andre Kolmogorov

$$\sum_{t=0}^{\alpha} |v_t|^2 \le \sum_{t=0}^{\alpha} |w_t|^2$$

for $\alpha = 0,1,2,\cdots$. More particularly, $|v_0|^2 < |w_0|^2$ (resp. $|v_0|^2 = |w_0|^2$) if and only if the all-pass wavelet p_t is non-trivial (resp. trivial).

Wold decomposition

Fundamental to the theory of stationary stochastic processes is the *Wold decomposition* (Wold, 1938, Theorem 7; Kolmogorov, 1941, Theorem 18). The following version can be derived from Theorem 7.70 in Robinson (1959, p. 97).

Theorem 8. x_t is a regular stationary stochastic process if and only if

$$x_t = \sum_{s=0}^{\infty} b_s \epsilon_{t-s}$$

where

(a) ϵ_t is an orthonormal sequence of random variables,

(b) ϵ_t lies in the closed linear manifold spanned by x_s, $s < t$,

(c) $b_0 > 0$,

(d) $\sum_{t=0}^{\infty} |b_t|^2 < \infty$

This representation of x_t, called the *Wold decomposition*, is unique.

Theorem 9. Let x_t be a regular stationary stochastic process with spectral density function $\Phi(f)$. Then the coefficients $b_0, b_1, b_2, \cdots$ of the Wold decomposition and the coefficients $\beta_0, \beta_1, \beta_2, \cdots$ of the minimum-delay wavelet with gain $\sqrt{\Phi(f)}$ are Identical; i.e., $b_t = \beta_t$ for $t = 0,1,2,\cdots$

Proof. Because the wavelets b_t and β_t have the same gain $\sqrt{\Phi(f)}$, and because, by definition, β_t is minimum-delay, we may apply Theorem 1 to obtain

$$\sum_{t=0}^{\infty} b_t z^t = \mathcal{P}(z) \sum_{t=0}^{\infty} \beta_t z^t, \quad |z| < 1$$

From Theorem 4.3 in Doob (1953, p. 577), we have

$$b_0 = \exp\left[\int_{-0.5}^{0.5} \log\sqrt{\Phi(f)}\, df\right]$$

Hence $b_0 = \beta_0$. It then follows from Theorem 7 that $\mathcal{P}(z)$ is trivial, and in fact $\mathcal{P}(z) = \mathcal{P}(0) = 1$. Q. E. D. [The condition $\sum_{t=0}^{\infty} b_t z^t \neq 0$ for $|z| < 1$ given in Doob's Theorem 4.3 is redundant.]

Extremal properties

The following theorem exhibits the extremal properties of the Wold decomposition, or in other words, of the minimum-delay wavelet.

Theorem 10. Let w_t be a wavelet in the class of all wavelets with gain $M(f)$. Then each of the following conditions is necessary and sufficient that

$$w_t = c\,\beta_t \text{ for } t = 0,1,2,\cdots$$

where β_t is the minimum-delay wavelet for the gain $M(f)$ and c is a complex constant of modulus $|c| = 1$.

(a) The group-delay of w_t is minimum for $-0.5 < f < 0.5$.

(b) The modulus $|\mathcal{W}(z)|$ is a maximum for all z satisfying $|z| < 1$.

(c) The partial energy $\sum_0^\alpha |w_t|^2$ is a maximum for $\alpha = 0,1,2,\cdots$

(d) For the regular stationary stochastic process

$$x_t = \sum_{s=0}^{\infty} w_s\, \epsilon_{t-s}$$

(where ϵ_t is an orthonormal sequence of random variables), the least-squares linear prediction $\hat{x}_{t+\alpha}$ of $x_{t+\alpha}$, $\alpha > 0$, from the whole past x_s, $s \leq t$, is

$$\hat{x}_{t+\alpha} = \sum_{s=\alpha}^{\infty} w_s\, \epsilon_{t+\alpha-s}$$

The prediction error is

$$\hat{x}_{t+\alpha} - x_{t+\alpha} = \sum_{s=0}^{\alpha-1} w_s\, \epsilon_{t+\alpha-s}$$

The minimum mean-square prediction error is the partial energy

Chapter 7. Andre Kolmogorov

$$E\{|\hat{x}_{t+\alpha} - x_{t+\alpha}|^2\} = \sum_{s=0}^{\alpha-1} |w_t|^2$$

(e) For the regular stationary stochastic process

$$x_t = \sum_{s=0}^{\infty} w_s \epsilon_{t-s}$$

the closed linear manifold spanned by x_s, $s < t$, is the same as the closed linear manifold spanned by ϵ_s, $s < t$, for all integers t.

(f) The regular stationary stochastic process

$$x_t = \sum_{s=0}^{\infty} w_s \epsilon_{t-s}$$

has an autoregressive representation in the sense that there exists complex numbers a_{Ns} such that

$$\epsilon_t = \text{l.i.m.}_{N \to \infty} \sum_{s=0}^{N} a_{Ns} x_{t-s}$$

where

$$\sum_{s=0}^{N} a_{Ns} z^s \neq 0 \quad \text{for} \quad |z| < 1$$

(g) For the regular stationary stochastic process

$$x_t = \sum_{s=0}^{\infty} w_s \epsilon_{t-s}$$

the transfer function $K(f)$ of the least-squares linear predictor of $x_{t+\alpha}$, $\alpha > 0$, from the whole past $x_s, s < t$, is

$$K(f) = \frac{1}{W(f)} \sum_{t=0}^{\infty} w_{t+\alpha} e^{-2\pi i f t}$$

(h) The set $\{w_{t+s}, s = 0, -1, -2, \cdots\}$ is closed in $L^2(0, \infty)$.

Proof. Conditions (a), (b), and (c) follow from Theorems 5, 6, and 7, respectively and Theorem 1. Condition (d) follows from Theorem 9 and from the Kolmogorov solution of the prediction problem (see Robinson,

1959, Section 7.8, equations (7.84) and (7.86)) Conditions (e) and (f) follow from Theorems 8 and 9. Condition (g) follows from condition (d) by utilizing the Stone-Kolmogorov isomorphism (see Robinson, 1959, Sections 7.2 and 7.8, especially equation (7.85)). The closed linear manifold spanned by $x_s, s \leq 0$, is the same as the closed linear manifold spanned by $\epsilon_s, s \leq 0$, if and only if every

$$z = \sum_{t=0}^{\infty} c_t \, \epsilon_{-t} \quad \text{where} \quad \sum_{t=0}^{\infty} |c_t|^2 < \infty$$

such that

$$\mathbb{E}\{z\bar{x}_s\} = 0 \quad \text{for all} \quad s \leq 0$$

satisfies $c_t \equiv 0$ (for all integers t). But

$$\mathbb{E}\{z\bar{x}_s\} = \mathbb{E}\left\{\sum_{t=0}^{\infty} c_t \, \epsilon_{-t} \sum_{r=0}^{\infty} \bar{w}_r \, \bar{\epsilon}_{s-r}\right\}$$

$$= \sum_{t=0}^{\infty} \sum_{r=0}^{\infty} c_t \, \overline{w}_r \, \delta_{-t-s+r} = \sum_{r=0}^{\infty} c_t \, \overline{w}_{t+s}$$

Condition (h) now follows by noting that the set $\{w_{t+s}, s \leq 0\}$ is closed in $L^2(0, \infty)$. if and only if any wavelet c_t orthogonal to each member of the set identically vanishes. Q.E.D.

Linear prediction

The following theorems given conditions for explicit representations of the linear predictor.

Theorem 11. Let x_t be a regular stationary stochastic process with spectral density $\Phi(f)$. A necessary and sufficient condition that the transfer function $K(f)$ of the least squares linear predictor (for any prediction lead $\alpha > 0$) be expressible as a Fourier series

$$K(f) = \sum_{s=0}^{\infty} k_s \, e^{-2\pi i f s}$$

where k_s is a wavelet is that

Chapter 7. Andre Kolmogorov

$$\int_{-0.5}^{0.5} \frac{df}{\Phi(f)} < \infty \qquad (2)$$

Proof. Necessary condition. By hypothesis,

$$K(f) = \frac{\sum_{s=0}^{\infty} b_{s+\alpha} e^{-2\pi i f s}}{\sum_{s=0}^{\infty} b_s e^{-2\pi i f s}} = \sum_{s=0}^{\infty} k_s e^{-2\pi i f s}, \quad \sum_{s=0}^{\infty} |k_s|^2 < \infty$$

Let $\alpha = 1$. Under the Stone-Kolmogorov isomorphism,

$$\hat{x}_1 \leftrightarrow K(f)$$

(see equation at bottom of p 103 in Robinson, 1959). Hence

$$\hat{x}_0 = \hat{x}_1 U^{-1} \leftrightarrow K(f) e^{-2\pi i f}$$

(see p. 86 in Robinson, 1959). Also

$$\hat{x}_t \leftrightarrow e^{-2\pi i f t}$$

and

$$\vartheta_t \leftrightarrow \frac{e^{-2\pi i f t}}{B(f)}$$

(see equation (7.74) in Robinson, 1959, p. 100). Therefore

$$b_0 \vartheta_0 = x_0 - \hat{x}_0 \leftrightarrow \frac{b_0}{B(f)} = 1 - K(f) e^{-2\pi i f}$$

It then follows that

$$\frac{1}{B(f)} = \frac{1}{b_0} - \frac{k_0}{b_0} e^{-2\pi i f} - \frac{k_1}{b_0} e^{-2\pi i f 2} - \cdots$$

where

$$\int_{-0.5}^{0.5} \frac{df}{|B(f)|^2} = \frac{1}{b_0^2} \left[1 + \sum_{s=0}^{\infty} |k_s|^2 \right]$$

which converges because, by hypothesis,

$$\sum_{s=0}^{\infty} |k_s|^2 < \infty$$

Since $\Phi(f) = |B(f)|^2$, the necessary condition is established. Q.E.D.

Proof. Sufficient condition. Now using more explicit notation to indicate the prediction lead α, we denote the transfer function of the optimum linear predictor by

$$K_\alpha(f) = \frac{B_\alpha(f)}{B(f)} \quad \text{where} \quad B_\alpha(f) = \sum_{t=0}^{\infty} b_{t+\alpha} e^{-2\pi i f t}$$

Because $B_0/B = 1$, it follows that B_0/B belongs to $L^2(df)$. Suppose B_α/B belongs to $L^2(df)$. We note that $1/B$ belongs to $L^2(df)$ since

$$|B(f)|^2 = \Phi(f)$$

and, by hypothesis, $1/\Phi(f)$ is integrable. Hence the linear combination

$$e^{2\pi i f} \frac{B_\alpha(f)}{B(f)} - b_\alpha e^{2\pi i f} \frac{1}{B(f)} = \frac{B_{\alpha+1}(f)}{B(f)}$$

belongs to $L^2(df)$. Therefore, by finite induction

$$K_\alpha(f) = \frac{B_\alpha(f)}{B(f)}$$

belongs to $L^2(df)$ for $\alpha = 0, 1, 2, \cdots$. Since, by Theorem 9, $B(f)$ is minimum-delay, we have the canonical representation

$$B(z) = \exp\left\{\int_{-0.5}^{0.5} \frac{e^{2\pi i f} + z}{e^{2\pi i f} - z} \log|B(f)|\, df\right\}$$

Since $b_\alpha, b_{\alpha+1}, \cdots$ is a wavelet, we have the canonical representation

$$B_\alpha(z) = P(z)\exp\left\{\int_{-0.5}^{0.5} \frac{e^{2\pi i f} + z}{e^{2\pi i f} - z} \log|B_\alpha(f)|\, df\right\}$$

Combining the above two equations, we have

$$\frac{B_\alpha(z)}{B(z)} = P(z)\exp\left\{\int_{-0.5}^{0.5} \frac{e^{2\pi i f} + z}{e^{2\pi i f} - z} \log\left|\frac{B_\alpha(f)}{B(f)}\right| df\right\} \quad (3)$$

Because B_α/B belongs to $L^2(df)$, the gain $|B_\alpha/B|$ is quadratically integrable. Also

$$\log \frac{|B_\alpha(f)|}{|B(f)|} = \log|B_\alpha(f)| - \log|B(f)|$$

Chapter 7. Andre Kolmogorov

is integrable. Hence equation (3) is the canonical representation of $\mathcal{K}(z)$, and so k_t is a wavelet. In fact, k_t is given by

$$k_t = \sum_{s=0}^{t} b_{\alpha+s}\, a_{t-s} \tag{4}$$

where a_t is the wavelet corresponding to $1/B(f)$. Q.E.D.

Theorem 12. Let x_t be a regular stationary stochastic process with spectral density $\Phi(f)$. A necessary and sufficient condition that the least-squares linear prediction $\hat{x}_{t+\alpha}$ (for any prediction lead $\alpha > 0$) has the representation

$$\hat{x}_{t+\alpha} = \sum_{s=0}^{\infty} k_s\, x_{t-s}$$

in the sense that k_s is a wavelet and that the error of approximation

$$\hat{x}_{t+\alpha} - \sum_{s=0}^{N} k_s\, x_{t-s}$$

belongs to the linear manifold $\Theta(t - N - 1)$ whose intersection

$$\bigcap_{N=0}^{\infty} \Theta(t - N - 1)$$

is that condition (2) holds.

Proof. We let $\Theta(t)$ denote the closed linear manifold spanned by $\vartheta(s), s < t$. We have

$$\hat{x}_\alpha = \sum_{s=0}^{N} k_s\, x_{t-s} = \sum_{s=0}^{\infty} b_{\alpha+n}\, \vartheta_{-n} - \sum_{s=0}^{N} k_s \left[\sum_{r=0}^{\infty} b_r\, \vartheta_{-s-r}\right]$$

$$= \sum_{n=0}^{\infty} b_{\alpha+n}\, \vartheta_{-n} - \sum_{n=0}^{\infty} \left[\sum_{s=0}^{N} k_s\, b_{n-s}\right] \vartheta_{-n}$$

$$= \sum_{n=0}^{\infty} \left[b_{\alpha+n} - \sum_{s=0}^{N} k_s\, b_{n-s}\right] \vartheta_{-n}$$

$$= \sum_{n=N+1}^{\infty} \left[b_{\alpha+n} - \sum_{s=0}^{N} k_s\, b_{n-s}\right] \vartheta_{-n} \in \Theta(-N-1)$$

if and only if

$$b_{\alpha+N} = \sum_{s=0}^{N} k_s b_{N-s}, \quad N = 0,1,2,\cdots \quad (5)$$

But, by Theorem 11, k_s is a wavelet satisfying equation (5) if and only if condition (2) holds. Q.E.D.

If we define the inner product between two random variables y and z with finite second absolute moments by $(y, z) = \mathbb{E}\{y\bar{z}\}$ then the space of all such random variables is a Hilbert space, denoted by $L^2(dP)$. Consider all complex-valued functions $H(f)$ for which

$$\int_{-0.5}^{0.5} |H(f)|^2 \, \Phi(f) \, df < \infty$$

where $\Phi(f)$ is a spectral density function. If we define the inner product between two such functions $H(f)$ and $G(f)$ to be

$$\int_{-0.5}^{0.5} H(f)\overline{G(f)} \, \Phi(f) \, df < \infty$$

then the space of all such functions is a Hilbert space, denoted by $L^2[\Phi(f)df]$.

Theorem 13. Let x_t be a regular stationary stochastic process with spectral density $\Phi(f)$. Let condition (2) be satisfied, so that the wavelet k_t given by equation (4) represents the coefficients of the least-squares linear predictor for prediction lead $\alpha > 0$. Then each of the following conditions is necessary and sufficient that

$$\hat{x}_{t+\alpha} = \mathrm{l.i.m.}_{N \to \infty} \sum_{s=0}^{N} k_s \, x_{t-s} \quad \text{in space } L^2[dP]$$

(a) $\quad K(f) = \mathrm{l.i.m.}_{N \to \infty} \sum_{s=0}^{N} k_s \, e^{-2\pi i f s} \quad \text{in space } L^2[\Phi df]$

(b) $\quad b_{\alpha+n} = \mathrm{l.i.m.}_{N \to \infty} \sum_{s=0}^{N} k_s \, b_{n-s} \quad \text{in space } L^2[0, \infty]$

Proof. We have

$$\mathbb{E}\left\{\left|\hat{x}_\alpha - \sum_{s=0}^{N} k_s\, x_{-s}\right|^2\right\} = \int_{-0.5}^{0.5} \left|K(f) - \sum_{s=0}^{N} k_s\, e^{-2\pi i f s}\right|^2 \Phi(f)\,df$$

by the Stone-Kolmogorov isomorphism), which is

$$\int_{-0.5}^{0.5} \left|B_\alpha(f) - \sum_{s=0}^{N} k_s\, e^{-2\pi i f s}\, B(f)\right|^2 df = \sum_{n=0}^{\infty} \left|b_{\alpha+n} - \sum_{s=0}^{N} k_s\, b_{n-s}\right|^2$$

(by Bessel's equality). Q.E.D.

Sufficient conditions that

$$\sum_{s=0}^{N} k_s\, x_{t-s}$$

converge in the mean (in space $L^2(dP)$) to $\hat{x}_{t+\alpha}$ have been given by Wold (1938), Doob (1953), and Yaglom (1955). These sufficient conditions involve the uniform convergence of the Fourier series

$$\sum_{s=0}^{N} k_s\, e^{-2\pi i f s}$$

Another sufficient condition given by Wiener and Masani (1958) is that
$0 < C_1 < \Phi(f) < C_2 < \infty$

Condition (c) in the theorem below improves on the result of Wiener and Masani.

Theorem 14. Let x_t be a regular stationary stochastic process with spectral density $\Phi(f)$, Let $K(f)$ be the transfer function of the least-squares linear predictor for prediction lead $\alpha > 0$. Then each of the following conditions is sufficient that there is a wavelet k_t such that

$$\hat{x}_{t+\alpha} = \underset{N \to \infty}{\text{l.i.m.}} \sum_{s=0}^{N} k_s\, x_{t-s} \quad \text{in space } L^2[dP]$$

(a) $\sum_{s=0}^{N} k_s\, e^{-2\pi i f s}$ converges uniformly to $K(f)$.

(b) $K(f) = \sum_{s=0}^{\infty} k_s e^{-2\pi i f s}$ where $\sum_{s=0}^{\infty} |k_s| < \infty$

(c) $\int_{-0.5}^{0.5} \frac{df}{\Phi(f)} < \infty$ and $\Phi(f) < C$ where C is a positive costant

Proof.

(a). Uniform convergence implies convergence (in the mean) in space $L^2(df)$ so that k_s is a wavelet. From Theorem 11 it follows that condition (2) holds, and that k_s is given by equation (4). Uniform convergence implies convergence (in the mean) in space $L^2[\Phi(f)df]$. Hence by condition (a) of Theorem 13, the desired result follows. Q.E.D.

(b). Because

$$\left| K(f) - \sum_{s=0}^{N} k_s e^{-2\pi i f s} \right| = \left| \sum_{s=N+1}^{\infty} k_s e^{-2\pi i f s} \right| \leq \left| \sum_{s=N+1}^{\infty} |k_s| \right|$$

it follows that the partial sums converge uniformly to $K(f)$, so that condition (a) is satisfied. Q.E.D.

(c). Theorem 11 applies and gives the wavelet k_s of equation (4). Because

$$\int_{-0.5}^{0.5} \left| K(f) - \sum_{s=0}^{N} k_s e^{-2\pi i f s} \right|^2 \Phi(f) df \leq C \int_{-0.5}^{0.5} \left| K(f) - \sum_{s=0}^{N} k_s e^{-2\pi i f s} \right|^2 df$$

the partial sums converge (in the mean) to $K(f)$ in space $L^2[\Phi(f)df]$. The desired result follows from condition (a) of Theorem 13. Q.E.D.

In closing, we recall that Kolmogorov (1941, Theorem 24) proved that condition (2) is necessary and sufficient that x_t have least-square interpolation error η_t such that $\mathbb{E}\{|\eta_t|^2\} > 0$.

I should like to express my sincere thanks to Professor Herman Wold for his help and encouragement.

Institute of Statistics, Uppsala University

Пocrnynuna a ребаксцаuvo , 6.5.61

References

[1] J. L. Doob, *Stochastic processes*, Wiley, New York, 1953.

[2] U. Grenander and. M. Rosenblatt, *Statistical analysis of stationary time-series*, Wiley, New York, and Almqvist and Wicksell, Uppsala, 1957.

[3] K. Karhunen, Ober die Struktur stationarer zufalliger Funktionen, *Ark. Mathematik*, 1 (1949), 141-160.

[4] V. I. Krylov, On functions regular in a half-plane (Russian), *Mat. Sbornik*, 6 (48), (1939), 95-138. (English translation by D. V. Thampuran, University of Wisconsin).

[5] A. N Kolmogorov, Stationary Sequences in Hilbert space (Russian), *Bull. Moscow State University*, Math, 2 (1941). (Spanish translation in Trab. Estad, 4, 55 and 243).

[6] I. I. Privalov, Randeigenschaften analytischer Functionen (German translation from 1950 Russian edition) *VEB Deutscher Verlag der Wissenschaften,* Berlin, 1956.

[7] F. Riesz, Ober die Randwerte einer analytischen Function, *Math. Zeit,* 18 (1923), 87-95. .

[8] E. A. Robinson*, Infinitely Many Variates*. Griffin, London, 1959.

[9] V. I. Smirnov, Sur les valuers limites des fonctions regulieres a l'interieur d'un cercle, *Leningrad Phys-Math*, 2 (1928).

[10] G. Szego), Uber die Randwerte einer analytischen Function, *Math. Ann.*, 84 (1921), 232-244.

[11] N. Wiener and P. Masani, The prediction theory of multivariate stochastic processes, II. *Acta Math.* (Uppsala), 99 (1958), 93-137.

[12] H. .Wold. A study in the analysis of stationary time-series, Thesis, University of Stockholm (Second Edition, Almqvist and Wiksell, Uppsala, 1954), 1938.

[13] A. M. Yaglom, Introduction to the theory of stationary random functions (Russian), *Uspehi Mat. Nauk*, 7 (1952), 3-168. (English translation by D. V. Thampuran, University of Wisconsin).

[14] A. M. Yaglom, Correlation theory of processes with random stationary n-th increments (Russian), *Math. Sbornik*, 37 (79), (1955), 141-196; (English translation in A. M. S. Translations, 8 (Series 2), 1958).

Chapter 8. Digicon

Texas Instruments

The deconvolution portion of Robinson's 1954 thesis (GAG Report No. 7) was published as the paper, "Predictive Decomposition of Seismic Traces," *Geophysics*, 1957. In the silver anniversary issue of *Geophysics* in 1960, Robinson was cited as a Classic Author of *Geophysics* with the following citation about the paper: "An influence far exceeding the immediate gains directly attributable to their suggested applications."

Digital signal processing and deconvolution were waiting for the companies to make use of them. One was Texas Instruments, which was the parent company of Geophysical Service, Inc. (GSI).

TEXAS INSTRUMENTS
INCORPORATED

3609 BUFFALO SPEEDWAY HOUSTON 6, TEXAS
JACKSON 6-1411 • P.O. BOX 6027 • CABLE: HOULAB

INDUSTRIAL INSTRUMENTATION DIVISION

June 3, 1957

Professor Steven Simpson
Geology and Geophysics Department
Massachusetts Institute of Technology
Cambridge 39, Massachusetts

Dear Professor Simpson:

We are manufacturing several Time Domain Filters for seismic work and would like to enclose a copy of E. A. Robertson's article, "Predictive Decomposition of Time Series with Application to Seismic Exploration", M.I.T. G.A.G. Report No. 7, 12 July 1954, with the Manual. Mr. Mark Smith, of Geophysical Service Inc., tells me you may be able to supply twelve or so copies. If so, they would be most helpful.

Very truly yours,

JULIAN H. UNGER
Engineer

In his column *From the Other Side* in the October 2009 issue of *The Leading Edge*, Lee Lawyer writes the following:

Chapter 8. Digicon

I received this from Bill Schneider who is living about a mile high in Colorado while I swelter at near sea level. Where is the justice in that? Of course, in a few months he will be a mile high in deep snow and I will be enjoying beautiful weather (maybe).

> I enjoyed your recent piece on deconvolution. I was there with GSI during those exciting early digital revolution days and would like to share my recollections on events that transpired relating to the development of deconvolution. First let me say we all agree that Milo's MAE process was a classic paper on dereverbation of marine records, and, as you correctly stated, it was successfully applied on Lake Maracaibo data well in advance of the first digital applications. The MAE process, however, was a deterministic solution to the water-reverberation problem. It was implemented on magnetically recorded data as a three-point tape-delay line filter and was limited by design to attenuate only water reverberations. The more general solution to deconvolution (i.e. removal of the seismic wavelet) was made possible by digital processing of digital recorded or converted (analog-to-digital) data. (LCL interruption: I really liked the three-point tape-delay line filter because you could really get your hands on that one.)
> My involvement started in June 1961 when I joined GSI as a research geophysicist fresh out of MIT. The research group was headed by Mark Smith and included Milo Backus, John Burg, Freeman Gilbert, and other notables. The group was heavily involved in developing digital processing applications for the then GSI-Mobil-Texaco exclusive digital field system contract. It was an exciting time and to keep up with the rapid progress we were making in plucking low-hanging digital fruit, we held weekly lunch hour technical presentations. In particular, I remember a series John Burg was giving on prediction error filtering, a statistical method which could remove the seismic wavelet from the unpredictable reflection series with knowledge only of the seismic trace power spectrum, assuming the wavelet was

minimum phase. Well, when this concept was tried on ringing marine data, the results were spectacular. Needless to say GSI moved quickly to patent the process to protect its interests. During this same period, I was reading up on the GAG (Geophysical Analysis Group) reports which I had acquired at MIT. In report No. 7 (I believe), I discovered Enders Robinson's work on predictive filtering which in effect was identical to the aforementioned Burg development. I noted this to Burg and others, but John professed no knowledge of Enders' work or the GAG reports, and I truly believe he independently developed the concept albeit somewhat later. The commercial ramifications of all this were rather exciting as well. Soon after Robinson's public domain work was recognized within GSI, a group lead by Rudy Prince split off to form Digicon and enticed Enders to join them in offering deconvolution processing services to the exploration industry. Not surprisingly, law suits and counter suits were filed and ultimately settled out of court.

The Origins of Digicon

by Edward R. (Rudy) Prince, Jr.

From Digest, The Quarterly Newsletter of Digicon Geophysical Corp. Fall 1995 (SEG Issue)

THIRTY YEARS AGO, six bright young men, each with $334 in our pockets and a great deal of energy, left the careers we had established at Geophysical Service, Inc. (GSI) to form a company called Digital Consultants, Inc. Today, like that well-known drum-beating rabbit, that

Chapter 8. Digicon

same company is 'still going' with the same unstoppable energy as when we started it all those years ago, but it is now known as Digicon, Inc.

This was the beginning of the company, but why would the six of us each take **$334 (for 334 shares of stock in the new company)**, leave the comfortable corporate environment of GSI and start a risky business where our individual necks were on the line? The origins of Digicon lie more in the answer to that question than in any one event which can be singled out.

In the late 1940's and early 1950's, GSI was a member of the Geophysical Analysis Group at the Massachusetts Institute of Technology. This Group had been founded to explore applications in geophysics, utilizing theory which became declassified following World War II. In the late 1950's, GSI began to investigate digital seismic data acquisition and processing techniques and, by the early 1960's, had established itself as a leading company in the seismic industry. The six founder members of Digicon all held key positions at GSI during this period of explosive growth and great success, during which time the seismic industry was literally revolutionized. In 1965, we felt that a rare opportunity was open to us. A massive revolution was imminent, as the transition from analog to digital technology for both seismic data recording and processing gained momentum, and these six individuals believed that we had the necessary abilities, expertise and drive to exploit this revolution and become successful on our own. That opportunity became Digicon.

The six founders who departed from GSI at this time were Dave Brown, George Cloudy, Pat Poe, Rudy Prince, Bill Shell and Dave Steetle. I have arranged our names in alphabetical order, as we were all equal shareholders with our $334 stakes! These six formed the Board of Directors, with Dave Brown elected Chairman and President and the other five elected Vice-Presidents. The company was incorporated in Texas on August 2nd, 1965, and we all decided to move to Houston to begin operations. **Within the first month of existence, Enders Robinson, previously Chairman of the Geophysical Analysis Group at MIT, joined the company as an equal member and became Vice-President of Research.**

Dave Brown was Manager of Special Projects at GSI, responsible for all digital land and marine crews and all computer centers. Dave served as Chairman and President of Digicon for over five years before departing the company in November 1970. He now resides in Boulder, Colorado, handling personal investments.

George Cloudy was the supervisor of the first digital land crew at GSI and of the first digital evaluation crew working for Exxon, which completed thirty-two separate projects during its first year. He went on to supervise the activities of several land and marine crews, including data processing, for GSI, before becoming Vice-President of Operations at Digicon. Except for a five-year period in the 1970's, when George went exploring for oil and gas, he remained with the company until July 1994 and is now an explorationist in the Rockies.

Pat Poe had become the lead programmer for the Texas Instruments Automatic Computers (TIAC's) at GSI. He confounded programmers at Texaco and Mobil by staging the autocorrelation, deconvolution design and filter application on the TIAC, which was a neat feat considering there were only 228 memory locations – that's not 228K, just 228! Pat became Vice-President of Programming for Digicon and remained so until the Computer Systems Division was sold to become Cogniseis, where he is currently Director of Strategic Planning.

Rudy Prince was in research at GSI, performing the first deconvolution operations, and I then became an Area Geophysicist, concentrating on marketing digital geophysical technology. I was later assigned to all digital marine crews and their data processing. I became Vice-President of Computer Operations and, later, of Marine Operations at Digicon, before being elected Chairman and President in November 1970. I am now Vice-Chairman of Zydeco Exploration, and remain a member of Digicon's Board of Directors.

Bill Shell had extensive experience in field operations with several contractors before joining GSI, where he became a supervisor of data processing. At Digicon, Bill continued to supervise field operations and data processing, as well as interpretation activities, until his departure from the company in 1970. He has been involved with personal investments in Houston since then.

Dave Steetle was second-in-charge of engineering at GSI, but was always the first in hands-on engineering, making projects actually work! At Digicon he was Vice-President of Engineering and was responsible for creating computer systems for data input and display. He was also in charge of field crew instrumentation and completely designed the industry's first successful digital marine streamer. Dave now resides in Galveston, but continues to lend his considerable expertise to Digicon, building systems and solving engineering problems on a consultancy basis.

Enders Robinson's career would take several pages to describe, but suffice it to say that he was truly the father of modern data processing in geophysics. After leaving Digicon in 1970, Enders became a professor at several universities and presently holds the Maurice Ewing Chair at Columbia University in New York.

These were the founding individuals of the company and it was the digital technology revolution which became our motivation for creating Digicon. In the thirty years of the company's existence since that time, there have been many momentous events I can recall, representing the successes and failures which have shaped the company as it is today. Beginning with the lawsuit filed against us by GSI during our first week in operation (later settled in our favor!); the involvement of Scientific Data Systems and the SDS 9300 computers; the special help extended to us in the early years by PetroFina and Amax; our relationship with Olympic (now Grant) Geophysical in the Gulf of Alaska; the incident with a US Navy plane, a heat-seeking missile and the M/V Pacific Seal; recording the first marine data north of Prudhoe and wintering the vessel in the Mackenzie Delta; the loss of the M/V Atlantic Seal in a fire on the east coast. I could go on, but this article is only about the beginnings of Digicon, so these other stories will have to wait until another time, or perhaps until future issues of Digest?

Letter from Pat Poe 1965

```
                              Digital Consultants Inc.
                              Westwick Building
                              3810 Westheimer
                              Houston, Texas

                              September 10, 1965
```

Dr. Enders A. Robinson
60 Belknap St.
Concord, Massachussetts

Dear Dr. Robinson,

This is to affirm our understanding of your position with Digital Consultants, Inc.

At a board of directors meeting on September 2, 1965, you were elected a vice-president and a member of the board of directors of Digital Consultants, Inc. The secretary was authorized to issue you 334 shares of stock.

We realize that you presently have a personal part time, independent research contract with Pan American Petroleum Corporation. We understand that, under no circumstances, will you divulge to us the nature or the findings obtained as a result of that contract until such time as you obtain publishing rights on these findings. We further understand that we should not contract your services in any way to conflict with the Pan American job.

Very truly yours,

Patrick H. Poe

Patrick H. Poe
Vice-president

Originally geophysical exploration was divided into two activities; namely, acquisition and interpretation. Now there were three; namely, acquisition, digital signal processing, and interpretation. Virtually all of the oil discovered in the past fifty years has been found by use of digital signal processing. Today DSP directs the steerable drill bit as it makes its way in horizontal drilling in the search for oil and natural gas in source rock as deep as four miles in the earth. The following gives a brief history of Digicon.

A history written by the geophysical company CGG

1965. Digital Consultants Inc. is founded in Houston by six engineers and geophysicists who share the vision of bringing evolving digital

Chapter 8. Digicon

computing technology to the geophysical industry – a new concept. Initial capitalization was from the six founding partners putting up just $334 each! The Company carries out its first data interpretation job in the North Sea.

1966. Digital Consultants install a state of the art SDS-9300 computer that allows multi-trace, multi-task programming without tape output. The first marine project of Digital Consultants' was QC work onboard two wooden-hulled marine seismic vessels in the North Sea, where WWII mines were still drifting!

1967. Digital Consultants deploys its first land seismic crew.

1970. Now with approximately 300 employees worldwide, Digicon opens its first overseas data processing center in Singapore.

1971. Digicon opens a second overseas data processing center in East Grinstead, UK. Digital Consultants reincorporates as Digicon Inc. and goes public on the American Stock Exchange.

1974. Digicon conducts extensive surveys in Alaska.

1979. Digicon deploys the geophysical industry's first ever digital marine seismic streamer - the DSS-240. Digicon becomes the first geophysical company to offer commercial depth migration, a quantum leap in seismic imaging fidelity.

1978. Veritas purchases FPS (Floating Point Systems) CPUs to replace Array Processors, achieving substantial increase in processing speed.

1980. Digicon launches its first custom-built seismic survey vessel - the 290-ton Digicon Explorer.

1981. Digicon employee John Sherwood invents DMO (Dip Move Out) data processing technique.

1983. Digicon opens a new data processing center in Brisbane, Australia –a significant expansion into new Asia Pacific markets.

1986. The increasing use of 3D seismic leads to a race to develop systems that offer an integrated package of data acquisition, processing and interpretation.

1987. Digicon records its first non-exclusive 3D marine data library program (in Mobile Bay, Gulf of Mexico).

1988. Digicon launches Massively Parallel Processing (MPP)initiative, including development of new Seismic TANGO data processing system to replace DISCO.

1994. Digicon becomes the first geophysical company to offer pre-stack time migration (3D MOVES), another breakthrough in seismic imaging at that time.

1996. Digicon and Veritas combine to form Veritas DGC Inc. (Note: DGC stands for Digcon.) This merger immediately upgrades its asset base, installing new HP and SUN computer systems and an NEC SX-4 supercomputer to enhance data processing capabilities.

1998. Veritas DGC installs industry-first, new-generation Data Visualization Center in Houston, USA. CGG carries out offshore surveys in the Gulf of Mexico. Veritas DGC vessel, the SR/V Veritas Viking, sets a record by towing the industry's first 12,000-meter streamer.

1999. Veritas DGC earns reputation for accomplishing difficult jungle projects.

2000. Veritas DGC opens new headquarters building in Houston, USA and installs three more Data Visualization Centers in Crawley, UK, Calgary, Canada, and Perth, Australia.

2005. Veritas DGC builds new Global Processing Facility (powered by 64-bit AMD Opteron™ dual-core processors) in Houston, doubling its seismic data processing power.

2007. CGG and Veritas DGC combine to create CGGVeritas, a leading global geophysical services and equipment company.

2010. Launch of BroadSeis seismic technology for recording broader bandwidths and maximizing, to a greater degree than ever before, the spatial and temporal resolution of seismic data.

2013. CGGVeritas acquires Fugro's Geoscience Division and becomes CGG, a fully integrated Geoscience company with a total workforce of approximately 10,000 staff working in over 70 locations.

Chapter 9. A Historical Perspective of Spectrum Estimation

John Tukey, Claude Shannon, Hendrik Bode

In the fall semester of 1955, I was invited to visit the Bell Telephone Laboratories at Murray Hill, New Jersey. The entire telephone network was analog. They were interested in my work on digital signal processing. I came with lots of computed examples of digital seismic processing. I conversed mostly with John Tukey, Claude Shannon, and Hendrik Wade Bode.

John Tukey was a half-time professor at Princeton and half-time at Bell Telephone Laboratories. Professor Tukey, a strong mathematician, was a practical and systematic scientist who was in the process of molding the mathematical discipline of statistics into a vision of his own making. He was to be admired. To a large extent he achieved this feat. Tukey had been interested in my work since I first corresponded with him in 1951. He was a kind person who always freely gave his help and encouragement. Professor Tukey envisioned the fast Fourier transform, which he published with Cooley in 1965.

Claude Shannon was interested in the practical side of digital signal processing. He studied my copious pages of computer print-outs and examined the graphs. No one before had ever had looked at the results in the intense way that he did. I can still hear him softly saying, "All that data; all that data." He emitted an aura of the vast and unbounded reach of the intellect. Here was someone who appreciated deconvolution for what it was. Shannon, the father of information theory, clearly understood the trials and tribulations involved.

Hendrik Wade Bode was a pioneer of modern control theory and electronic telecommunications. His book *Network Analysis and Feedback Amplifier Design* is classic in the field of electronic

Chapter 9. A Historical Perspective of Spectrum Estimation

telecommunications. All of his work was with analog systems. He introduced the plot of asymptotic magnitude and phase that bears his name, the Bode plot. These plots display the frequency response of systems. He introduced the concept of minimum-phase.

I spent the evening at Bode's house. His wife prepared the dinner for the three of us. They were very gracious. He was especially interested in the theoretical side of digital signal processing. He applauded my extension of the minimum-phase concept from analog systems to digital signal processing. He appreciated my conversion of the frequency-domain concept of **minimum phase** to the time-domain concept of **minimum delay**.). Both terms mean precisely the same thing.

In 1984, I received the **Donald G. Fink Prize Award** given for the outstanding survey, review or tutorial paper in any of the Institute of Electrical and Electronic Engineers (IEEE) Transactions, Journals, Magazines, or Proceedings and awarded for "A Historical Perspective of Spectrum Estimation," *Proceedings of the IEEE*, vol. 70, pp. 885-907,

1982. The paper follows. It deals with things as they were when the paper was published in 1982.

Spectral estimation

The prehistory of spectral estimation has its roots in ancient times with the development of the calendar and the clock. The work of Pythagoras in 600 B.C. on the laws of musical harmony found mathematical expression in the eighteenth century in terms of the wave equation. The struggle to understand the solution of the wave equation was filially resolved by Jean Baptiste Joseph de Fourier in 1807 with his introduction of the Fourier series. The Fourier theory was extended to the case of arbitrary orthogonal functions by Sturm and Liouville in 1836. The Sturm-Liouville theory led to the greatest empirical success of spectral analysis yet obtained, namely the formulation of quantum mechanics as given by Heisenberg and Schrodinger in 1925 and 1926. In 1929 John von Neumann put the spectral theory of the atom on a firm mathematical foundation in his spectral representation theorem in Hilbert space. Meanwhile, Wiener developed the mathematical theory of Brownian movement in 1923, and in 1930 he introduced generalized harmonic analysis, that is, the spectral representation of a stationary random process. The common ground of the spectral representations of von Neumann and Wiener is the Hilbert space; the von Neumann result is for a Hermitian operator, whereas the Wiener result is for a unitary operator. Thus these two spectral representations are related by the Cayley-Mobius transformation. In 1942 Wiener applied his methods to problems of prediction and filtering, and his work was interpreted and extended by Norman Levinson. Wiener in his empirical work put more emphasis on the autocorrelation function than on the power spectrum.

The modern history of spectral estimation begins with the breakthrough of J. W. Tukey in 1949 on *Measuring Noise Color*, which is the statistical counterpart of the breakthrough of Fourier 142 years earlier. This result made possible an active development of empirical spectral analysis by research workers in all scientific disciplines. However, spectral analysis was computationally expensive. A major computational breakthrough occurred with the publication in 1965 of the fast Fourier transform

Chapter 9. A Historical Perspective of Spectrum Estimation

algorithm by J. S. Cooley and J. W. Tukey. The Cooley-Tukey method made it practical to do signal processing on waveforms in either the time or the frequency domain, something never practical with continuous systems. The Fourier transform became not just a theoretical description, but a tool. With the development of the fast Fourier transform the field of empirical spectral analysis grew from obscurity to importance, and is now a major discipline. Further important contributions were the development of spectral windows by Emmanuel Parzen and others starting in the 1950's, the statistical work of Maurice Priestley and his school, hypothesis testing in time series analysis by Peter Whittle starting in 1951, the Box-Jenkins approach by George Box and G. M. Jenkins in 1970, and autoregressive spectral estimation and order-determining criteria by E. Parzen and H. Akaike starting in the 1960's. To these statistical contributions must be added the equally important engineering contributions to empirical spectrum analysis, which are not treated at all in this paper, but form the subject matter of the other papers in this special issue.

Spectral estimation has its roots in ancient times, with the determination of the length of the day, the phases of the moon, and the length of the year. The calendar and the clock resulted from empirical spectral analysis. In modern times, credit for the empirical discovery of spectra goes to the diversified genius of Sir Isaac Newton [1]. But the great interest in spectral analysis made its appearance only a little more than a century ago. The prominent German chemist Robert Wilhelm Bunsen (1811-1899) repeated Newton's experiment of the glass prism. Only Bunsen did not use the sun's rays as Newton did. Newton had found that a ray of sunlight is expanded into a band of many colors, the spectrum of the rainbow. In Bunsen's experiment, the role of pure sunlight was replaced by the burning of an old rag that had been soaked in a salt solution (sodium chloride). The beautiful rainbow of Newton did not appear. The spectrum, which Bunsen saw, only exhibited a few narrow lines, nothing more. One of the lines was a bright yellow.

Bunsen conveyed this result to Gustav Robert Kirchhoff (1824-1887), another well-known German scientist. They knew that the role of the glass prism consisted only in sorting the incident rays of light into their respective wavelengths (the process known as dispersion). The Newton rainbow was the extended continuous band of the solar spectrum; it indicates that all wavelengths of visible light are present in pure sunlight. The yellow line, which appeared when the light source was a burning rag, indicated that the spectrum of table salt contained a single specific wavelength. Further experiments showed that this yellow line belonged to the element sodium. No matter what the substance in which sodium appeared, that element made its whereabouts known by its bright yellow spectral line. As time went on, it was found that every chemical element has its own characteristic spectrum, and that the spectrum of a given element is always the same, no matter in what compound or substance the element is found. Thus the spectrum identifies the element, and in this way we can tell what elements are in substances from the distant stars to microscopic objects.

The successes of spectral analysis were colossal. However, the spectral theory of the elements could not be explained by classical physics. As we know, quantum physics was born and spectral theory was explained in 1925 and 1926 by the work of Werner Heisenberg (1901-1976) and Erwin Schrodinger (1887-1961). In this paper, we will show how spectral theory developed in the path to this great achievement.

Although most of the glamour of spectral theory has been associated with quantum physics, we will not neglect the parallel path taken in classical physics. Although the two paths began diverging with the work of Charles Sturm (1803-1855) and Joseph Lionville (1809-1882) on the spectral theory of differential equations, we will see that the final results, namely, the spectral representation of John von Neumann (1903-1957) for quantum physics, and that of Norbert Wiener (1894-1964) for classical physics, are intimately related.

Because light has high frequencies, our instruments cannot respond fast enough to directly measure the waveforms. Instead, the instruments measure the amount of energy in the frequency bands of interest. The measurement and analysis of the spectra of other types of signals, however, take different forms. With lower frequency signals, such as mechanical vibrations, speech, sonar signals, seismic traces, cardiograms, stock market data, and so on, we can measure the signals as functions of time (that is, as time series) and then find the spectra by computation. With the advent of the digital computer, numerical spectrum estimation has become an important field of study.

Let us say a few words about the terms "spectrum" and "spectral." Sir Isaac Newton introduced the scientific term "spectrum" using the Latin word for an image. Today, in English, we have the word specter meaning ghost or apparition, and the corresponding adjective spectral. We also have the scientific word spectrum and the dictionary lists the word spectral as the corresponding adjective. Thus "spectral" has two meanings. Some feel that we should be careful to use "spectrum" in place of "spectral" a) whenever the reference is to data or physical phenomena, and b) whenever the word modifies "estimation." They feel that the word "spectral," with its unnecessary ghostly interpretations, should be confined to those usages in a mathematical discipline where the term is deeply embedded.

The material in this paper through the section "Wiener-Levinson Prediction Theory" surveying the period from antiquity through Wiener and Levinson can be described as "The Prehistory of Spectrum Estimation" to emphasize that spectrum estimation is interpreted as estimation from data. The remaining sections may be described as "Some Pioneering Contributions to the Development of Methods of Spectrum Estimation."

Modern spectrum estimation began with the breakthrough for the analysis of short time series made by J. W. Tukey in 1949. This work led to a great blossoming forth of spectrum analysis techniques. Despite the advances in digital computing machinery, such computations were still expensive. The next great breakthrough occurred with the discovery of the fast Fourier transform in 1965 independently by J. W. Cooley and J. W. Tukey and by Gordon Sande. This development, in conjunction with silicon chip technology, has brought spectrum analysis to bear on a wide range of problems.

Taylor series

At the time when calculus was introduced in the seventeenth century by Newton and Leibnitz, the concept of a mathematical "function" entailed restricted properties, which in the course of time were gradually made less severe. In those days, the observations of natural events seemed to indicate that continuous relations always existed between physical variables. This view was reinforced by the formulation of the laws of nature on the basis of differential equations, as exemplified by Newton's laws. Thus it became commonplace to assume that any function describing physical phenomena would be differentiable. The idea of a function that changes in some capricious or random way, and thus does not allow any analytic formula for its representation, did not enter into the thinking of the mathematicians of that time. It was thus very natural for Brook Taylor (1685-1731) [2], a contemporary of Newton, to introduce the concept of "analytic function." The Taylor series expands an analytic function as an infinite summation of component functions. More precisely, the Taylor series expands a function $f(x)$, which is analytic in the neighborhood of a certain point $x = a$, into an infinite series whose coefficients are the successive derivatives of the function at the given point

$$f(a+h) = f(a) + \frac{1}{1!} f'(a) h + \frac{1}{2!} f''(a) h^2 + \cdots$$

Thus analytic functions are functions which can be differentiated to any degree. We know that the definition of the derivative of any order at the point $x = a$ does not require more than knowledge of the function in an arbitrarily small neighborhood of the point $x = a$. The astonishing property of the Taylor series is that the shape of the function at a finite distance h from the point $x = a$ is uniquely determined by the behavior of the function in the infinitesimal vicinity of the point $x = a$. Thus the Taylor series implies that an analytic function has a very strong interconnected structure; by studying the function in a small vicinity of the point $x = a$, we can precisely predict what happens at the point $x = a + h$, which is at a finite distance from the point of study. This property, however, is restricted to the class of analytic functions. The best known analytic functions are, of course, the sine and cosine functions, the polynomials, and the rational functions (away from their poles).

Chapter 9. A Historical Perspective of Spectrum Estimation

Daniel Bernoulli solution of the wave equation

The great Greek mathematician Pythagoras (c 570 B.C.-c 495 B.C.) was the first to consider a purely physical problem in which spectrum analysis made its appearance. Pythagoras studied the laws of musical harmony by generating pure sine vibrations on a vibrating string, fixed at its two endpoints. This problem excited scientists since ancient days, but the mathematical turning point came in the eighteenth century when it was recognized that the vertical displacement $u(x, t)$ of the vibrating string satisfies the wave equation

$$\frac{\partial^2 u}{\partial x^2} - \frac{1}{c^2}\frac{\partial^2 u}{\partial t^2} = 0$$

Here x is the horizontal coordinate and t is the time. The constant c is a physical quantity characteristic of the material of the string, and represents the velocity of the traveling waves on the string. Because the endpoints $x = 0$ and $x = \pi$ are fixed, we have the boundary conditions

$$u(0, t) = u(\pi, t) = 0$$

(Note that for simplicity, we have taken the string to be of length π.) The problem of constructing the solution of the wave equation was then attacked by some of the greatest mathematicians of all time, and in so doing, they paved the way for the theory of spectrum analysis.

One of the finest results was that of Daniel Bernoulli (1700-1782) [3] in 1738. He introduced the method of separation of variables in which a trial solution is constructed as the product of a function of x alone, and a function of t alone. Thus, he wrote

$$u(x, t) = X(x)\, T(t)$$

Putting this trial solution in the differential equation and solving, he found the solutions

$$\cos kx \cos kct, \quad \cos kx \sin kct$$
$$\sin kx \cos kct, \quad \sin kx \sin kct$$

However, the boundary condition at $x = 0$ excludes the solutions involving $\cos kx$, and so the possible solutions are reduced to the two choices

$$\sin kx \cos kct, \quad \sin kx \sin kct$$

The boundary condition at $x = \pi$ requires that the value of k to be an integer. In view of the linearity of the wave equation, any superposition of solutions gives a solution. Bernoulli thus gave the following solution:

$$u(x,t) = \sum_{k=1}^{\infty} \sin kx \, (A_k \cos kct + B_k \sin kct)$$

where the A_k and B_k are arbitrary constants. Bernoulli made the claim that this infinite sum is the general solution of the equation for the vibrating string. The implications of Bernoulli's claim were startling.

From the principles of mechanics it was known that the initial displacement and initial velocity of the string could be prescribed in an arbitrary way. That is, it was known that at the initial time $t = 0$, both $u(x,0)$ and $\dot{u}(x,0)$ could have any functional form. (Note that the dot over a function indicates differentiation with respect to time, so $\dot{u}$ represents the velocity of the string in the vertical direction.) However, Bernoulli's solution gives explicit expressions for initial displacement and initial velocity, namely,

$$u(x,0) = \sum_{k=1}^{\infty} A_k \sin kx$$

$$\dot{u}(x,0) = c \sum_{k=1}^{\infty} A_k \sin kx$$

Thus Bernoulli's solution implied that each of two arbitrary functions $u(x,0)$ and $\dot{u}(x,0)$ could be expanded in the interval $0 \leq x \leq \pi$ in the form of an infinite series of sine functions. However, this result could not be explicitly demonstrated in Bernoulli's time.

Bernoulli's result can be expressed in the following way. Let the initial displacement $u(x,0)$ be an arbitrary nonanalytic function $f(x)$. Then we have the expansion

$$f(x) = \sum_{k=0}^{\infty} A_k \sin kx$$

which says that a nonanalytic function $f(x)$ can be expressed as an infinite summation of analytic functions $\sin kx$ with weighting

Chapter 9. A Historical Perspective of Spectrum Estimation

coefficients A_k. This result was a paradox at the time, and it led to a historical controversy of whether the function $f(x)$ could be freely chosen or must be restricted to the class of analytic functions. From the physical point of view, $f(x)$, which is the initial displacement of the string, could be freely chosen. From the then contemporary mathematical point of view, $f(x)$, which is an infinite summation of analytic functions, must be analytic. This view was believed by all the eminent mathematicians of the day.

Two of the greatest mathematicians who ever lived then set out to find the coefficients A_k of this expansion. Multiply each side by $\sin nx$ and integrate between 0 and π. Because

$$\int_0^\pi \sin kx \, \sin nx \, dx = \begin{cases} \dfrac{\pi}{2} & \text{when } k = n \\ 0 & \text{when } k \neq n \end{cases}$$

the result found by L. Euler (1707-1783) [4] and J. L. Lagrange (1736-1813) [5] is

$$A_n = \frac{2}{\pi} \int_0^\pi f(x) \sin nx \, dx$$

This is the point at which the question stood at the start of the nineteenth century.

Jean Baptiste Joseph de Fourier and the sinusoidal spectral theory

On December 21, 1807 the engineer Jean Baptiste Joseph de Fourier (1768-1830) [6] addressed the French Academy and made a claim that appeared incredible to the eminent mathematicians who were members of the Academy. As it turned out, one of the greatest advances in the history of mathematics, an innovation which was to occupy much of the attention of the mathematical community for over a century, was made by an engineer. Fourier said at that historic meeting that an arbitrary function, defined over a finite interval by any rough and even discontinuous graph, could be represented as an infinite summation of cosine and sine functions. The distinguished and brilliant academicians questioned the validity of Fourier's theorem, for they believed that any superposition of cosine and sine functions could only give an analytic function, that is, an infinitely differentiable function. An analytic function, of course, could never be discontinuous, and thus was very far removed from some arbitrarily drawn graph.

In fact, Taylor's theorem stated that an analytic function had the property that, given its shape in an infinitesimal interval, the continuation of its course to the right and left by finite amounts was uniquely determined (the so-called process of analytic continuation). The academicians and the other great mathematicians of the time could not reconcile the property of analytic continuation with Fourier's theorem. How could the physical reasoning of an engineer stand up against the weight of the analytic reasoning of some of the most eminent mathematicians of all time? These were the days when many great men were at the peak of their powers. Yet Fourier stood alone in defending his theorem.

As we have seen, the concept of analytic function requires a strong interconnection of the values of a function, where knowledge at one point allows us to predict the value at a point at a finite distance h. This prediction mechanism is embodied in the Taylor series expansion. However, a non-analytic function, such as a rough and discontinuous function, does not demand any such prediction mechanism between the immediate vicinity of a point and its wider surroundings. The Fourier series expansion is stated in terms of this wider concept of function. The coefficients of a Fourier series, as shown by the Euler-Lagrange result, are obtained by integration and not by differentiation as in the case of the Taylor series. Each Fourier coefficient A_n is obtained by integrating $f(x) \sin nx$ over the entire range. Thus any modification of $f(x)$ in a limited portion of the range changes all of the Fourier coefficients. It follows that the interconnections operate in the Fourier series in a global sense and not in a local sense as in the case of the Taylor series. It is the behavior of $f(x)$ in the large that matters in the case of the Fourier series, and not so much the behavior in the vicinity of a point. How can we resolve the differences between these two types of expansions: the Taylor series, which is the expansion about a point which gives strict predictions a finite distance from the point, and the

Chapter 9. A Historical Perspective of Spectrum Estimation

Fourier series, which is an expansion in the large and which gives knowledge of the function in the entire range. The Taylor series requires unlimited differentiability at a point, whereas the Fourier series does not demand any differentiability properties whatever.

Surprisingly enough, the chasm between the Taylor series and the Fourier series is bridged by means of the z-transform, which is the fundamental transform used in the theory of digital signal processing. Let us consider an analytic function $f(z)$ of the complex variable

$$z = x + iy$$

We now expand the function in a Taylor series (in the variable z^{-1}) about the point $z^{-1} = 0$, to obtain the z-transform

$$f(z) = \sum_{n=0}^{\infty} a_n z^{-n}$$

The radius of convergence of this series extends from $z^{-1} = 0$ to the first singular point, say, z_0^{-1}. A singular point is a point where the function ceases to be analytic. The region of convergence of the Taylor series expansion of $f(z)$ is the region in the z-plane outside the circle of radius $|z_0|$; that is, the region of convergence is for all points z such that $|z^{-1}| < |z_0^{-1}|$ or equivalently $|z| > |z_0|$.

Let us now write the Taylor series expansion for points on the unit circle

$$z = x + iy = e^{i\theta} = \cos\theta + i\sin\theta$$

We have

$$f(z) = \sum_{n=0}^{\infty} a_n z^{-n} = \sum_{n=0}^{\infty} a_n e^{-ni\theta} = \sum_{n=0}^{\infty} a_n (\cos n\theta - i \sin n\theta)$$

which is in the form of a complex Fourier series in the angle θ. Three cases can occur.

In the first case, the singular point z_0 is inside the unit circle in the z-plane. In this case, the function is analytic on the unit circle and the Fourier series thus is an analytic representation of this analytic function. The French Academy believed this was the only case.

In the second case, the singular point z_0 is outside the unit circle. In this case, the Taylor series does not represent the function, and so we will not consider the case further.

The third case is the interesting one, and is the case which resolves the mathematical controversy which led up to Fourier's discovery in 1807. When the singular point z_0 lies on the unit circle, the Taylor series will not converge at some or all of the points on the unit circle. Thus the Taylor series defines an analytic function, which is differentiable to any order outside the unit circle, but the function becomes nonanalytic at some or all of the points on the unit circle. The Fourier series in θ is the Taylor series for z on the unit circle, and thus the Fourier series represents a function in the variable θ, which is nonanalytic at some or all of the points in its range $-\pi \leq \theta \leq \pi$. A small modification of the Fourier coefficients that would move the singular point z_0 from on the unit circle to just inside the unit circle would change a nonanalytic Fourier representation to an analytic one. The amazing thing is that it is enough to move the singularity from the periphery of the unit circle to the inside by an arbitrarily small amount, in order to change the given non-differentiable function in θ to one which can be differentiated any number of times. Thus the mistake of the great French mathematicians of the prestigious French Academy who wanted to restrict the validity of Fourier series to analytic functions depended entirely on that extremely small but finite distance from a point on the periphery to a point just inside the unit circle. A function can be extremely smooth right up to the unit circle, and then disintegrate into a rough and distorted image of its former self once it is on the unit circle. The Taylor series breaks down on the unit circle, but its counterpart, the Fourier series in θ, is still valid. The theorem of Fourier is true; science could blossom.

Sturm-Liouville spectral theory of differential equations

Following the great innovation of Fourier in 1807, the remarkable properties of Fourier series were gradually developed throughout the nineteenth century and into the twentieth century. The Fourier series as introduced by Fourier is an expansion in terms of cosines and sines,

Chapter 9. A Historical Perspective of Spectrum Estimation

which represent an orthogonal set of functions. However, there are many other sets of orthogonal functions, and so today any such expansion in terms of orthogonal functions is called a Fourier series. As we will see, some sets of orthogonal functions can be stochastic, and it turns out that the corresponding Fourier series play an important role in statistical spectral analysis.

First, however, let us look at the important generalizations made by the French mathematicians Charles Sturm (1803-1855) [7] and Joseph Liouville (1809-1882) [8] in the decade of the 1830's. Let us now briefly look at the Sturm-Liouville theory of differential equations. The vibration of any infinitely long right circular cylinder of radius one can be described by a second-order differential equation. Let us consider a simple case, namely, the differential equation (the one-dimensional Helmholtz equation)

$$u''(x) + k^2 u(x) = 0$$

The Helmholtz equation can be obtained by taking the temporal Fourier transform of the wave equation, which set off the search for the theory of Fourier. Here k is the wavenumber which is equal to ω/c where ω is the temporal frequency. In the Helmholtz equation, k^2 is some undetermined parameter. The variable x is the central angle of the cylinder, and so x lies in the range $-\pi$ and π. Because the points $x = -\pi$ and $x = \pi$ represent the same point on the cylinder, we must have the two boundary conditions

$$u(-\pi) = u(\pi)$$
$$u'(-\pi) = u'(\pi)$$

The general solution of the differential equation is

$$u(x) = A \cos kx + B \sin kx$$

The two boundary conditions restrict the choice of the parameter k^2 to the discrete set of values

$$k^2 = 0, 1^2, 2^2, 3^2, \cdots$$

which are called the eigenvalues of the Helmholtz equation. The corresponding solutions of the equation, namely, the functions

$$u_k(x) = A \cos kx + B \sin kx$$

are called the eigenfunctions. These eigenfunctions are the cosine and sine functions which Fourier had used to construct his Fourier series. These functions represent the characteristic vibrational modes of the cylinder, which can only vibrate in this discrete set of wavenumbers $k = 0, 1, 2, \cdots$. Thus the Sturm-Liouville theory has given the answer to why the discrete set of cosine and sine functions was the correct one for Fourier to use in a problem which stemmed from the wave equation.

Furthermore, the Sturm-Liouville theory gives us added insight to spectral analysis and, in fact, is the foundation of the spectral theory of differential equations. Most of the eigenvalue problems of mathematical physics are characterized by differential operators H of the form

$$H = \frac{d}{dx}\left[A(x)\frac{d}{dx}\right] + B(x)$$

The physical problems we consider require that the function $A(x)$ be positive within the given interval. Let us now form the operation $vHu - uHv$, which is

$$vHu - uHv = \frac{d}{dx}[A(x)(vu' - uv')]$$

We notice that the right-hand side is a total derivative, and so we have

$$\int_a^b (vHu - uHv)dx = [A(x)(vu' - uv')]_a^b$$

Any differential operator H, which allows the transformation of such an integral (as on the left) into a pure boundary term (as on the right), is called self-adjoint. Thus the Sturm-Liouville operator H is self-adjoint. Often we may prescribe boundary conditions so that the right-hand side vanishes; such boundary conditions are called self-adjoint. We then have a self-adjoint problem, namely, a problem characterized by a self-adjoint operator H and self-adjoint boundary conditions. We then have the identity in the functions $u(x)$ and $v(x)$ given by

$$\int_a^b (vHu - uHv)dx = 0$$

which is called Green's identity.

Chapter 9. A Historical Perspective of Spectrum Estimation

The eigenvalue problem associated with the self-adjoint operator H starts with the differential equation

$$H\phi = \lambda\phi$$

A solution satisfying the boundary conditions does not exist for all values of λ, but only for a certain selected set λ_i called the eigenvalues. This set consists of an infinite number of eigenvalues λ_i which are all real and which tend to infinity with i. We generally arrange these eigenvalues in increasing order to obtain the infinite sequence (called the spectrum)

$$\lambda_1, \lambda_2, \lambda_3, \cdots$$

together with the corresponding eigenfunctions

$$\phi_1, \phi_2, \phi_3, \cdots$$

We now consider two different eigenvalues λ_j, λ_k and their corresponding eigenfunctions ϕ_j, ϕ_k. If we substitute $u = \phi_j$ and $v = \phi_k$ into Green's identity, we obtain

$$\int_a^b (\lambda_j \phi_j \phi_k - \lambda_k \phi_k \phi_j)\, dx = 0$$

which gives the orthogonality condition

$$\int_a^b \phi_j(x) \phi_k(x)\, dx = 0, \qquad \text{for } j \neq k$$

By normalization, we can require that

$$\int_a^b \phi_j^2(x)\, dx = 0$$

so that the eigen-functions form an orthonormal set. The orthonormal property can be written more concisely as

$$\int_a^b \phi_j(x) \phi_k(x)\, dx = \delta_{jk}$$

where δ_{jk} is the Kronecker delta function.

Let us now represent an arbitrary function $f(x)$ in the form of the infinite expansion

$$f(x) = \sum_{k=1}^{\infty} c_k \phi_k(x)$$

As we have previously mentioned, such an expansion is called a *Fourier series* in honor of the pioneering work of Fourier. The Fourier coefficients c_k are obtained by multiplying both sides by $\phi_j(x)$ and integrating. The result is

$$c_j = \int_a^b f(x) \phi_j(x) \, dx$$

Under certain general conditions, it can be shown that the orthonormal set is complete, so that the above Fourier expansion actually converges to the function $f(x)$. Suppose now that $f(x)$ is the solution to the inhomogeneous differential equation

$$Hf(x) = p(x)$$

In terms of linear system theory, $p(x)$ is the input and $f(x)$ is the output. Now substitute $u = f$ and $v = \phi_k$ into Green's identity. We obtain

$$\int_a^b (\phi_k Hf - fH\phi_k) \, dx = 0$$

which is

$$\int_a^b (\phi_k p - f\lambda_k \phi_k) \, dx = 0$$

The above equation can be written as

$$\int_a^b f\phi_k \, dx = \frac{1}{\lambda_k} \int_a^b \phi_k p \, dx$$

We recognize the left-hand side as the expression for the Fourier coefficient c_k. Thus

$$c_k = \frac{1}{\lambda_k} \int_a^b \phi_k(\xi) p(\xi) \, d\xi$$

We now substitute this expression for c_k into the Fourier series to obtain

Chapter 9. A Historical Perspective of Spectrum Estimation

$$f(x) = \sum_{k=1}^{\infty} c_k \phi_k(x) = \int_a^b p(\xi) \left[\sum_{k=1}^{\infty} \frac{\phi_k(x)\phi_k(\xi)}{\lambda_k} \right] d\xi$$

If we denote the expression in brackets by $G(x,\xi)$, then this equation is

$$f(x) = \int_a^b p(\xi) G(x,\xi)\, d\xi$$

This is the integral form of the input-output relationship, and we recognize

$$G(x,\xi) = \sum_{k=1}^{\infty} \frac{\phi_k(x)\phi_k(\xi)}{\lambda_k}$$

as the *impulse response function* or *Green's function* (under the given boundary conditions), a concept originated by George Green (1793-1841) [9]. This equation exhibits the impulse response function of a linear system in terms of its spectrum $\lambda_1, \lambda_2, \lambda_3, \cdots$. We can confirm that the Green's function is indeed the impulse response by setting the input $p(x)$ equal to the impulse $\delta(x - x_0)$. Then the output is

$$\int_a^b \delta(\xi - x_0) G(x,\xi)\, d\xi = G(x, x_0)$$

and so $G(x, x_0)$ represents the output at x due to an impulse at x_0. Since the differential equation represents an input-output system, we see that the Green's function satisfies

$$H G(x, x_0) = \delta(x - x_0)$$

This equation shows that the Green's function $G(x, x_0)$ is the inverse of the differential operator H.

We have thus reviewed the spectral theory of differential operators, and now we can look at the most spectacular application of spectral estimation—quantum physics.

Schrodinger spectral theory of the atom

The Sturm-Liouville theory of the expansion of functions in terms of orthogonal functions found numerous physical applications in the work of Lord Rayleigh (1842-1919). Such expansions occur throughout the study of the elastic vibrations of solids and in the theory of sound. In the history of physics, a decisive breakthrough occurred when Erwin Schrodinger (1887-1961) [10] showed in 1926 that the vibrations occurring within the atom can be understood by means of the Sturm-Liouville theory. Let us now explain how the wave mechanics of Schrodinger describes the spectral lines of the atom. An equivalent matrix mechanics was formulated a year before Schrodinger by Werner Heisenberg (1901-1976) [11].

Before quantum theory, classical physics was at an impasse. It could not explain the existence of atomic spectra. For example, the bright yellow spectral line of sodium discovered by Bunsen means that the radiation of its atoms produces a discrete frequency ω_0. If we assume that this line is emitted by an electron, then the laws of classical physics state that such an electron should emit not a discrete line at ω_0, but a whole spectrum of lines at all frequencies ω, and with no discontinuities in the spectrum. That is, classical physics predicts that the spectrum of an electron should be continuous as is the spectrum of the sun. Yet Bunsen observed the discrete spectrum of sodium as evidenced by the bright yellow line. (As we will soon see, this line observed by Bunsen is actually a doublet, which Bunsen was unable to resolve with the means available to him.)

Quantum mechanics allows us to see the atom from a new point of view. Quantum mechanics says that atomic electrons jump from one energy state to another, and that the difference of these energies is embodied as a quantum of electromagnetic energy, the photon. If the energy diminishes, a photon is born. If the energy increases, a photon or a quantum of energy from some other field has been absorbed just before the jump.

In quantum mechanics, an electron is represented by a probability density function. (The probability density function is found as the squared magnitude $|\phi|^2$ of a probability wave function ϕ.) An electron jump has a probability that depends upon the shapes of the probability density functions that correspond to the electron prior to and after the jump. The probability of a jump is, generally speaking, greater for the stronger overlapping or deeper interpenetration of these probability density functions. The laws that divide electron transitions in atoms into more probable and less probable ones are called selection rules. It is in this jumping of electrons that photons are born. These photons enter a spectroscope, get sorted into types, and produce the spectral lines.

The more photons that an atom emits in a second, the brighter the spectral lines. If the number of atoms remains constant, then the brightness of the spectral lines depends upon the statistical frequency of electron jumps in the atoms. And this statistical frequency is determined by the probability distribution of jumps. It is in this way that an atomic spectrum consisting of a number of lines of different brightnesses is generated.

One can make the observation that the spectrum estimation problem (the subject matter of this special issue of *Proceedings of the IEEE*) is not central to the spectral representation in quantum mechanics. This situation was brought forcibly to the writer's attention several years ago at the U.S. Air Force Geophysics Library at Hanscom Field, MA, which is one of the best scientific libraries in the world. The many shelves devoted to "spectra" consisted of a mixture of both kinds of books, but no book devoted to a discussion of the relationship between the two areas of spectral theory.

Spectral estimation in quantum mechanics is based on the edifice of spectroscopy, which is an instrumentational science. In 1891, the physicist A. A. Michelson developed an interferometer, a device producing the superposition of a light signal on top of itself with a prescribed delay. In one series of experiments, Michelson first bandpass filtered a light signal by passing it through a prism. He then used the interferometer to measure the visibility of the superimposed signal as a function of delay. The resulting curve was the autocovariance function of the original signal. Michelson then used a mechanical harmonic analyzer to compute the Fourier transform of the visibility curve; that is, he estimated the power spectrum of the signal. Michelson's experiments were done to examine the fine structure of spectral lines of light. Thus in those early days, the present day dichotomy of spectrum estimation had not yet materialized.

The technique of spectral analysis in physics developed rapidly in the twentieth century, and the instruments became more powerful and sensitive. The spectroscopists came up with the following question for theoreticians, namely, the question of why spectral lines are somewhat fat, not infinitesimally thin.

It was recognized that a photon corresponds to a line at one frequency ω. The question was why the lines on a photographic plate of a spectroscope come out somewhat broadened, not slender. The answer was found in the wave property of the electron and the Heisenberg uncertainty principle. The initial energy of an electron in an atom refers to a stationary state, and so does the final energy. However, an electron jump is in violation of some steady state. As soon as this occurs, the Heisenberg principle takes over. If we let Δt designate the lifetime of an electron between jumps, then the uncertainty of photon energy is $\Delta E \sim h/\Delta t$, where h is Planck's constant. Using Planck's formula for energy quanta, the uncertainty ΔE of the energy is proportional to the uncertainty $\Delta \omega$ of the frequency of the photon

$$\Delta E = \frac{h}{2\pi} \Delta \omega$$

Thus the spectral lines have a width $\Delta \omega$ which is inversely proportional to the time of the "settled life" of the electron in the atom

$$\Delta \omega \sim \frac{2\pi}{\Delta t}$$

In other words, the more "settled" or quiescent the life of the electron in the atom, the narrower the spectral lines. That is why at high temperatures and pressures, when many of the atomic electrons are unsettled, the spectral lines broaden out and become smeared. Thus an individual spectral line has a finite width associated with thermal motion and collision broadening. This is not only important in physics, but it relates very importantly to the topic of spectrum estimation in this special issue. Real "lines" have finite width. This means that real lines behave like narrow-band noises and not like either single frequencies or a constant-amplitude lightly frequency-modulated signal.

Chapter 9. A Historical Perspective of Spectrum Estimation

Let us now return the discussion of the yellow sodium line which Bunsen observed. The sodium D line is a doublet. Moreover the sodium spectrum contains four lines in the visible range, and two more in the near ultraviolet, strong enough to be useful for analytic chemistry. The sodium spectrum contains 29 lines of astrophysical interest between the D lines and 4390 Å (still in the visible).

We might say that Bunsen over a century ago was performing spectrum estimation. He was unable to resolve the two frequencies present in the doublet, even as today a person doing spectrum estimation might have the same problem in some other situation. Also Bunsen missed the many other lines in the sodium atom, even as today a person doing spectrum estimation might not find some features without the use of modern techniques. As spectroscopic instruments became better, these lines were discovered. Now another question, however, has come up. Many spectral lines, which, it would seem, should correspond to a single frequency, actually turned out to be the states of a number of very close-lying lines. The fact that the sodium D line is a doublet is a case in point. The fine structures of spectral lines (doublets, etc.) were revealed only because of the great advances in spectral techniques. In turn, electron spin was discovered in order to explain these "fine qualities" in spectra. Let us briefly give the reason. When spectra are generated, the states of two electrons with opposite spins can have slightly different energies. As a result, the spectral line is doubled; in place of one line we have twin lines with identical brightnesses. Such twins are usually born only when the outer electron shell has one electron. If the number of electrons in this shell increases, we can have triplets and even larger families of the former spectral line.

Let us now consider the quantum mechanical formulation of the harmonic oscillator problem. In terms of the non-dimensional displacement x, the time-independent *Schrodinger equation* is

$$H\phi = \lambda\phi$$

where H is defined as the differential operator

$$H = \frac{d^2}{dx^2} - x^2$$

and λ is defined as

$$\lambda = \frac{2E}{\hbar \omega_0}$$

Here ϕ is the probability wave function, the constant E is the energy, $h = 2\pi\hbar$ is Planck's constant, and the constant ω_0 is the natural frequency. The problem of finding the probability wave function ϕ is a Sturm-Liouville problem. The solution gives the eigenvalues as $1, 3, 5, 7, \cdots$, and so we write

$$\lambda_k = (2k+1) \text{ for } k = 0, 1, 2, \cdots$$

Thus the eigenenergies are

$$E_k = \frac{1}{2}\hbar\omega_0 \lambda_k = \hbar\omega_0 \left(k + \frac{1}{2}\right) \text{ for } k = 0, 1, 2, \cdots$$

The corresponding eigenfunctions are

$$\phi_k = C_k h_k(x) e^{-x^2/2} \text{ for } k = 0, 1, 2, \cdots$$

where C_k is a normalization constant, and $h_k(x)$ is the Hermite polynomial of order k. The discrete set of eigenenergies $E_0, E_1, E_2, \cdots$ represent the discrete lines observed in the spectrum. Thus quantum mechanics, through the use of Sturm-Liouville theory, is able to explain the existence of atomic spectra. However, certain mathematical difficulties remained; the history of their resolution is given in the next section.

The von Neumann spectral representation theorem

In finite-dimensional space, the following eigenvalue problem is posed. Given an Hermitian matrix H, find all column-vector solutions ϕ of the characteristic equation

$$H\phi = \lambda\phi$$

where λ is a constant also to be determined. That is, given H, find ϕ and λ. The solutions $\phi_1, \cdots, \phi_n$ are called the eigensolutions (assumed to be normalized), and the corresponding real numbers $\lambda_1, \cdots, \lambda_n$ are called the eigenvalues of the matrix H. The totality of the eigenvalues $\lambda_1, \lambda_2, \cdots, \lambda_n$, in order of increasing magnitude, is called the spectrum. Now write the eigen-equations

Chapter 9. A Historical Perspective of Spectrum Estimation

$$H\phi_k = \lambda \phi_k \quad \text{for } k = 1, 2, \cdots n$$

in the form of the matrix equation

$$HU = U\Lambda$$

Because the eigensolutions are orthonormal, the matrix U (which has the eigensolutions as it columns) is unitary, i.e.,

$$UU^T = I$$

where I is the identity matrix. (The superscript T indicates complex conjugate transpose.) The matrix Λ is diagonal matrix, with the spectrum along its diagonal. Thus this eigenvalue problem can be described as the problem of finding a unitary matrix U that reduces H to a real diagonal matrix,

$$U^{-1}HU = \Lambda$$

(Note: In case H is real, then H is a symmetric matrix and U is an orthogonal matrix.)

Although the unitary matrix U, whose columns are the eigensolutions ϕ_i, is not uniquely determined by H, John von Neumann [12] in 1929 exploited the unitary nature of U to reformulate the eigenvalue problem. The von Neumann reformulation, which is called the *spectral representation problem*, yields the same results as the eigenvalue problem in finite-dimensional space, but has the advantage that it can be extended to Hilbert space.

We recall that the diagonal matrix Λ is defined to be the matrix with the eigenvalues, ordered by increasing magnitude, along its main diagonal and zeroes off the diagonal. Because of this ordering, the matrix Λ is uniquely determined for any given Hermitian matrix H. Because some eigenvalues may be repeated, let us relabel them as $\lambda_1, \lambda_2, \cdots, \lambda_m$ (with $m \leq n$), where each λ_i is now distinct. Consequently for a given H, we have the unique decomposition

$$\Lambda = \lambda_1 Q_1 + \lambda_2 Q_2 + \cdots + \lambda_m Q_m$$

where Q_i is a diagonal matrix with 1's in those places on its main diagonal in which λ_i occurs in Λ and 0's elsewhere. The sum of the Q_i gives the identity matrix

$$I = Q_1 + Q_2 + \cdots + Q_m$$

We now define the matrix P_j as

$$P_j = UQ_jU^{-1} \text{ (for } j = 1, 2, \cdots, m)$$

A projection matrix is defined as a Hermitian idempotent matrix. Because Q_j is Hermitian ($Q_j = Q_j^T$) and idempotent ($Q_jQ_j = Q_j$), it follows that Q_j is a projection matrix. Because P_j is Hermitian ($P_j = P_j^T$) and idempotent

$$P_jP_j = UQ_jU^{-1}UQ_jU^{-1} = UQ_jQ_jU^{-1} = UQ_jU^{-1} = P_j$$

it follows that P_j is a projection matrix. Since for $i \neq j$

$$P_iP_j = UQ_iU^{-1}UQ_jU^{-1} = 0$$

it follows that $P_i + P_j$ is a projection matrix and the space spanned by P_i is orthogonal to the space spanned by P_j. Let us now define the function $\mathcal{H}(\lambda)$ of the continuous variable λ as

$$\mathcal{H}(\lambda) = P_1\delta(\lambda - \lambda_1) + P_2\delta(\lambda - \lambda_2) + \cdots + P_m\delta(\lambda - \lambda_m)$$

This function is the continuous representation of the suite of projection matrices $P_1, P_2, \cdots, P_m$.

We now consider the quadratic form uHv where u is a row vector and v is a column vector. We have

$$uHv = uU\Lambda U^{-1}v = uU(\lambda_1Q_1 + \lambda_2Q_2 + \cdots + \lambda_mQ_m)U^{-1}v$$
$$= u(\lambda_1P_1 + \lambda_2P_2 + \cdots + \lambda_mP_m)v$$
$$= \lambda_1uP_1v + \lambda_2uP_2v + \cdots + \lambda_muP_mv$$

The essence of the von Neumann spectral representation lies in the fact that the components uP_jv are numerically invariant for given u, H, and v. In this way, the nonuniqueness of the unitary matrix U appearing in the eigenvalue decomposition is bypassed. We see that we can write the quadratic form as the integral

$$uHv = \int_{-\infty}^{\infty} \lambda\, u\, \mathcal{H}(\lambda)\, v\, d\lambda$$

This equation represents the von Neumann *spectral representation of the Hermitian matrix* H.

Let us now analyze this equation. If we strip the u and v from this equation, we are left with

Chapter 9. A Historical Perspective of Spectrum Estimation

$$H = \int_{-\infty}^{\infty} \lambda\, \mathcal{H}(\lambda)\, d\lambda$$

which, in matrix notation, is

$$H = \lambda_1 P_1 + \lambda_2 P_2 + \cdots + \lambda_m P_m$$

We can write the row vector u as

$$u = \int_{-\infty}^{\infty} u\, \mathcal{H}(\lambda)\, d\lambda$$

which is

$$u = u P_1 + u P_2 + \cdots + u P_m$$

Finally, we can write

$$Hv = \int_{-\infty}^{\infty} \lambda\, \mathcal{H}(\lambda)\, v\, d\lambda$$

which is

$$Hv = \lambda_1 P_1 v + \lambda_2 P_2 v + \cdots + \lambda_m P_m v$$

Let us now consider functions of the matrix H. First, we consider the square of H. We have

$$H^2 = (\lambda_1 P_1 + \cdots + \lambda_m P_m)^2 = \lambda_1^2 P_1 + \cdots + \lambda_m^2 P_m = \int_{-\infty}^{\infty} \lambda^2\, \mathcal{H}(\lambda)\, d\lambda$$

We see that squaring H results in squaring the λ inside the integral. In general, if we form a function of H, then the result is that the same function of λ is taken within the integral sign; that is

$$f(H) = \int_{-\infty}^{\infty} f(\lambda)\, \mathcal{H}(\lambda)\, d\lambda$$

The above spectral representation was derived for finite-dimensional space, that is, a space in which the elements u and v are vectors and the Hermitian operator H is a matrix. One of the major achievements of von Neumann was the development of the concept of the infinitely-dimensional space, which he called *Hilbert space* in honor of the great mathematician David Hilbert (1862-1943). We now let u and v represent elements in Hilbert space, and let H represent a Hermitian operator. A Hilbert space is characterized by an inner product (or dot product). The inner product of the elements u and v is denoted by

$\langle u, v \rangle$. If we let H operate on the element v, we obtain a new element Hv. The inner product of the elements u and Hv is denoted by $\langle u, Hv \rangle$.

This inner product is the counterpart of the quadratic form uHv in finite-dimensional space. Once we establish this connection, it turns out that the von Neumann spectral representation has exactly the same form in Hilbert space as it does in finite-dimensional space. Thus in Hilbert space, we also have an operator $\mathcal{H}(\lambda)$, which is the continuous representation of the suite of projection operators associated with the Hermitian operator H. Whereas in finite-dimensional space, we made use of the quadratic form $u\mathcal{H}(\lambda)v$, we now make use of its counterpart $\langle u, \mathcal{H}(\lambda)v \rangle$ in Hilbert space. Thus the von Neumann *spectral representation* in Hilbert space is

$$\langle u, \mathcal{H}v \rangle = \int_{-\infty}^{\infty} \lambda \, (u, \mathcal{H}(\lambda) \, v) \, d\lambda$$

Let us now look at some history. In general, there is no quadratically integrable solution to the eigenvalue problem in Hilbert space. This circumstance, however, bothered no one working in physics. Wavelet solutions (i.e., quadratically integrable superpositions of eigenfunctions with eigenvalues in a small neighborhood) were used from the start, appearing in the works of de Broglie and Schrodinger from 1924.

One of the authors cited in the Reference Section knew von Neumann personally, studied his work assiduously, and certainly regards him as one of the truly great founders of quantum theory. However, there was never a "crisis in physics" that was resolved by the von Neumann spectral representation theorem. Most people doing the practical calculations to be compared with experiment had never heard of the theorem, which was for them at such a high level of abstraction that it had no bearing on what they were doing.

Throughout this essay we have traced the development of spectral theory, from the analytic functions of Brook Taylor, to the non-differentiable functions of Jean Baptiste Joseph de Fourier, and now to the more general operators of Hilbert space. At each stage, these developments were mathematical in nature, but they laid the foundations for subsequent advances in physics. Reasoning in

Chapter 9. A Historical Perspective of Spectrum Estimation

mathematics and reasoning in physics often appear quite different. When a major physical breakthrough occurs, such as in quantum mechanics in the 1920's, and a flood of exciting new physical results come out, certainly the work of mathematicians in establishing existence and uniqueness theorems might seem somewhat irrelevant.

For a moment let us go back to Six Isaac Newton. It is often said that the unique greatness of Newton's mind and work consists in the combination of a supreme experimental with a supreme mathematical genius. It is also often said that the distinctive feature of Newtonian science consists precisely in the linking together of mathematics and experiment, that is, in the mathematical treatment of experimental or (as in astronomy, geophysics, or wherever experiments cannot be performed) observational data. Yet, although correct, this description does not seem to be quite complete; there is more in the work of Newton than mathematics and experiment. There is also a deep intuition and insight in his interpretation of nature.

In today's science, specialization has gone far. Physicists use mathematics; they formulate problems, devise methods of solution, and perform long derivations and calculations, but generally they are not interested in creating new mathematics. The discovery and purification of abstract concepts and principles is particularly in the realm of mathematics. John von Neumann (1903-1957) is a prime example of a mathematician doing physics. When he did physics, he thought and calculated like a physicist, only faster.. He understood all branches of physics, as well as chemistry and astronomy, but mainly he had a talent for introducing only those mathematical ideas that were relevant to the physics at hand. The introduction of abstract Hilbert space theory in quantum mechanics, chiefly by von Neumann, made possible the construction of a solid theory on the basis of the powerful intuitive ideas of Dirac and other physicists.

The physics of quantum theory cannot be mathematically formulated in finite-dimensional space but requires Hilbert space. After the work of Heisenberg and Schrodinger in 1925 and 1926, there was a crisis in abstract mathematics because the physics of quantum mechanics could not be adequately formulated in terms of the existing mathematical framework. This situation was rectified in 1929 by von Neumann [12] who laid the mathematical foundations of quantum mechanics in terms of Hilbert space. There is an apocryphal story that the young John von Neumann, who was barely past being a teenager, and had not yet earned his doctorate, was lecturing in Gottingen. Of course, most of the famous physicists present regarded his work as too abstract, but the great mathematician Hilbert was in the audience. As the story goes, the elderly Hilbert leaned over and whispered into Professor Courant's ear: "What is this Hilbert space?" Another even more apocryphal story goes as follows. A group of physicists came to von Neumann and described a problem in physics which they could not solve. After thinking for a while, von Neumann in his head came up with the numerical answer which agreed with the experimental result, which the physicists knew but had not told him. They were very impressed and they blurted out "Dr. von Neumann, the general solution involves solving an infinite set of nonlinear partial differential equations. Certainly you have found some mathematical shortcut!" von Neumann answered "No, I solved the infinite set."

von Neumann [13] showed that from a mathematical point of view, it is the spectral representation that is required in quantum mechanics rather than the solution of the eigenvalue problem as such. In this sense, spectral theory represents the key to the understanding of the atom. In fact, von Neumann [13] has shown that the spectral representation enters so essentially into all quantum mechanical concepts that its existence cannot be dispensed with. His establishment of the spectral representation of the Hermitian operator H is one of the great achievements in mathematics, and a milestone in the history of spectral theory.

Einstein-Wiener theory of Brownian motion

A highly interesting kinetic phenomenon known as Brownian movement was first reported in 1827 by the distinguished botanist, Robert Brown, who found that "extremely minute particles of solid matter when suspended in pure water exhibit motions for which I am unable to account and which, from their irregularity and seeming independence,

Chapter 9. A Historical Perspective of Spectrum Estimation

resemble in a remarkable degree, the less rapid motions of some of the simplest animalcules of infusions." This type of irregular zigzag movement is typified by the dancing of dust particles in a beam of light. The cause of Brownian movement was long in doubt, but with the development of the kinetic theory of matter came the realization that the particles move because they are bombarded unequally on different sides by the rapidly moving molecules of the fluid in which they are suspended. The Brownian movement never ceases. The detailed physical theory of Brownian movement was worked out in 1904 by M. von Smoluchowski [14], and in a more final form in 1905 by Albert Einstein [15]. In 1923, Norbert Wiener [16] developed the mathematical theory of Brownian movement, which today is the basis of the mathematical model of white noise in continuous time. White noise is defined as a stationary random process which has a constant spectral power density. The concept of the white noise process, as given by the Einstein-Wiener theory of Brownian motion, is important in all theoretical studies of spectrum analysis.

In practice, a signal is of finite duration, and usually can be digitized on a grid fine enough for interpolation to be adequate. In this sense, the set of data representing a signal is really finite. Accordingly, we do not have to go to continuous time or to infinite time unless 1) we so wish or 2) we gain from it. In other words, as long as we stay finite, we do not need the Einstein-Weiner theory. With this *caveat emptor*, let us now discuss this theory.

A white noise process in continuous time cannot be represented by the ordinary types of mathematical functions which one meets in calculus. Instead, white noise can only be represented by what mathematicians call a *generalized function*. The most familiar example of generalized function is the *Dirac delta function*, which is often defined as

$$\delta(t - t_0) = 0 \text{ for } t \neq t_0$$
$$\delta(t - t_0) = \infty \text{ for } t = t_0$$
$$\int_{-\infty}^{\infty} \delta(t - t_0) dt = 1$$

The most important property of the delta function is its sifting property, that is, its ability to isolate or reproduce a particular value of an ordinary function $f(t)$ according to the convolution formula

$$\int_{-\infty}^{\infty} f(t-t_0)\,\delta(t)dt = f(t_0)$$

If one feels uncomfortable with generalized functions, then one can often avoid them by using Lebesgue-Stieltjes integrals. For example, the Heaviside step function $H(t)$ is an ordinary function equal to zero for $t < 0$ and to one for $t \geq 0$. Since

$$dH(t) = \delta(t)dt$$

the above convolution formula becomes the Lebesgue-Stieltjes integral

$$\int_{-\infty}^{\infty} f(t-t_0)\,dH(t) = f(t_0)$$

This Lebesgue-Stieltjes integral involves only ordinary functions.

Let us now look at a *white noise process* which we denote by $\epsilon(t)$. It is a generalized random function. Again let $f(t)$ be an ordinary function, and consider the convolution integral

$$\int_{-\infty}^{\infty} f(t-t_0)\,\epsilon(t)dt$$

Let $\mathcal{E}(t)$ be the integrated white noise process, so that we may write

$$d\mathcal{E}(t) = \epsilon(t)\,dt$$

The integrated white noise process $\mathcal{E}(t)$ is an ordinary random function, and the above convolution becomes the Legesgue-Stieltjes integral

$$\int_{-\infty}^{\infty} f(t-t_0)\,d\mathcal{E}(t)$$

Wiener formulated everything in terms of Lebesgue-Stieltjes integrals with ordinary functions. However, we are going to take a strictly engineering approach and formulate things in terms of ordinary integrals, but with generalized functions.

Without loss of generality in the discussion which follows, we can for convenience let $t_0 = 0$, so that the integral in question becomes

Chapter 9. A Historical Perspective of Spectrum Estimation

$$\int_{-\infty}^{\infty} f(t)\,\epsilon(t)\,dt$$

As is usual statistical practice, let $\mathbb{E}$ denote the mathematical expectation operator. Since this operator is linear, it may be interchanged with integral signs (provided certain regularity conditions hold). The expectation of the above integral is

$$\mathbb{E}\int_{-\infty}^{\infty} f(t)\,\epsilon(t)\,dt = \int_{-\infty}^{\infty} f(t)\,\mathbb{E}\epsilon(t)\,dt$$

Because we want white noise to have zero mean, we let $\mathbb{E}\epsilon(t) = 0$, and so the above integral is zero. Let us next consider the variance given by

$$\mathbb{E}\left[\int_{-\infty}^{\infty} f(t)\,\epsilon(t)\,dt\right]^2$$

$$= \mathbb{E}\int_{-\infty}^{\infty} f(t)\,\epsilon(t)\,dt \int_{-\infty}^{\infty} f(\tau)\,\epsilon(\tau)\,d\tau$$

$$= \int_{-\infty}^{\infty}\int_{-\infty}^{\infty} f(t)\,f(\tau)\,\mathbb{E}[\epsilon(t)\,\epsilon(\tau)]\,dt\,d\tau$$

Now we come to the key point. We want white noise to be uncorrelated at two different time points, but at the same time we want the variance of white noise to produce an impulse so as to make the above integral have a nonzero value. Thus the key element is to define the covariance $\mathbb{E}[\epsilon(t)\epsilon(\tau)]$ as being equal to $\delta(t-\tau)$. Then the above integral becomes

$$\int_{-\infty}^{\infty}\int_{-\infty}^{\infty} f(t)\,f(\tau)\,\delta(t-\tau)\,dt\,d\tau = \int_{-\infty}^{\infty} f^2(t)\,dt$$

We can therefore make the following definition. A generalized random function $\epsilon(t)$ is white noise provided that

$$\mathbb{E}\epsilon(t) = 0 \text{ and } \mathbb{E}[\epsilon(t)\epsilon(\tau)] = \delta(t-\tau)$$

For a long time such a random process was regarded as improper. As we know, the delta function can be approximated arbitrarily close by ordinary functions. Likewise, the white noise process $\epsilon(t)$ can be approximated arbitrarily close by ordinary random processes.

Because one never uses the white noise process in isolation but only in integrals, the white noise process can be avoided by the use of the

Lebesgue-Stieltjes integral, just as the Dirac delta function can be so avoided. However, as we have said, we will not follow the Lebesgue-Stieltjes approach here.

Let us now consider white noise $\epsilon(n)$ for discrete (integer) time n. White noise in discrete time is not a generalized random process, for $\epsilon(n)$ is merely a sequence of zero-mean, constant-variance, uncorrelated random variables. However, the Fourier transform of discrete white noise is a generalized random process, which we denote by $E(\omega)$. We have

$$E(\omega) = \sum_{n=-\infty}^{\infty} \epsilon(n) e^{-i\omega n} \quad \text{for } -\pi \leq \omega \leq \pi$$

We can easily verify that $E(\omega)$ has zero mean. The covariance of $E(\omega)$ is

$$\mathbb{E}[E^*(\omega)E(\mu)] = \mathbb{E} \sum_{n=-\infty}^{\infty} \epsilon(n) e^{i\omega n} \sum_{k=-\infty}^{\infty} \epsilon(k) e^{-i\mu k}$$

$$= \sum_{n=-\infty}^{\infty} \sum_{k=-\infty}^{\infty} \mathbb{E}[\epsilon(n)\epsilon(k)] e^{i(\omega n - \mu k)}$$

Because $\mathbb{E}[\epsilon(n)\epsilon(k)] = \delta_{nk}$ (i.e., the Kronecker delta function, which is one when $n = k$ and zero otherwise), we have

$$\mathbb{E}[E^*(\omega)E(\mu)] = \sum_{n=-\infty}^{\infty} e^{-in(\mu-\omega)} = 2\pi \delta(\mu - \omega)$$

That is, the covariance is a Dirac delta function. Thus we come to an important result: The Fourier transform of a white noise process in (infinitely extended) discrete time n is a white noise process in the continuous variable ω. (It is easy to show that the corresponding result holds for the case of a white noise process in continuous time.) In other words, the Fourier transform of a very rough (white) process in time is a very rough (white) process in frequency. The Fourier transform preserves (saves) information and does not smooth (destroy) information. Today this result is second-nature to an engineer, but when Wiener obtained this result in 1923 it was startling. Wiener unlocked the spectral theory of the most random of processes (white noise), and now the stage was set for applying this result to the more smooth processes which are generated by many physical phenomena. Wiener made this application in 1930 under the name of *generalized harmonic analysis*, but before we give its history we will break our train of thought and look at the innovative work of Yule in 1927. Yule's work at the time seemed modest. While most mathematicians and physicists were developing general methods to deal with the infinite and the infinitesimal in spectrum analysis, Yule was developing a simple model with a finite number of parameters (i.e., a finite parameter model) in order to handle spectrum analysis in those cases where this model was appropriate. This model of Yule is known as the autoregressive (AR) process.

Yule autoregressive spectrum estimation method

At the turn of the twentieth century, Sir Arthur Schuster [17] introduced a numerical method of spectrum analysis for empirical time series. Let $x(n)$ represent the value of a time series at discrete (integer) time n. Given N observations of the time series from $n = 1$ to $n = N$, then Schuster's method consisted of computing the periodogram $P(\omega)$ defined as

$$P(\omega) = \frac{1}{N}\left[x(1)e^{-i\omega} + x(2)e^{-i\omega 2} + \cdots + x(N)e^{-i\omega N}\right]^2$$

For example, suppose that the time series consists of a sinusoid of frequency ω_0 with superposed errors; then, the periodogram would show a peak at $\omega = \omega_0$. Thus by computing the periodogram, the peaks would show the location of the frequencies of the underlying sinusoidal motion. Until the work of Yule (1871-1951) in 1927 [18], the Schuster periodogram approach was the only numerical method of empirical spectrum analysis. However, many empirical time series observed in nature yielded a periodogram that was very erratic and did not exhibit any dominant peaks. This led Yule to devise his autoregressive method of spectrum analysis. In those days, empirical spectrum analysis was called the investigation of periodicities in disturbed series. His main application was the determination of the spectrum of Wolfer's sunspot time series.

G. Udny Yule in 1927 introduced the concept of a finite parameter model for a stationary random process in his fundamental paper on the investigation of the periodicities in time series with special reference to Wolfer's sunspot numbers. If we consider a curve representing a sinusoidal function of time and superpose on the ordinate small random errors, then the only effect is to make the graph somewhat irregular, leaving the suggestion of periodicity still quite clear to the eye. If the errors are increased in magnitude, the graph becomes more irregular, the suggestion of periodicity more obscure, and we have only sufficiently to increase the errors to mask completely any appearance of periodicity. But, however large the errors, Schuster's periodogram analysis is applicable to such a time series, and given a sufficient number of observations should yield a close approximation to the period and amplitude of the underlying sinusoidal wave.

Yule reasoned in the following way. Consider a case in which periodogram analysis is applied to a time series generated by some physical phenomenon in the expectation of eliciting one or more true periodicities. Then it seemed to Yule that in such a case there would be a tendency to start with the initial hypothesis that the true periodicities are masked solely by additive random noise. As we well know, additive random noise does not in any way disturb the steady course of the underlying sinusoidal function or functions. It is true that the periodogram itself will indicate the truth or otherwise of the hypothesis made, but Yule saw no reason for assuming it to be the hypothesis most likely a priori.

At this point, Yule introduced the concept of an input-output feed-back model. The amplitude of a simple harmonic pendulum with damping (in discrete approximation) can be represented by the homogeneous difference equation

$$b(n) + a_1 b(n-1) + a_2 b(n-2) = 0$$

Here $b(n)$ is the amplitude at discrete (integer) time n. Errors of observation would cause superposed fluctuations on $b(n)$, but Yule observed that, by improvement of apparatus and automatic methods of recording, errors of observation can be practically eliminated: An initial

Chapter 9. A Historical Perspective of Spectrum Estimation

impulse or disturbance would set the pendulum in motion, and the solution of the difference equation would give the impulse response. The initial conditions are $b(n) = 0$ for $n < 0$ and $b(0) = 1$. The characteristic equation of the difference equation is

$$E^2 + a_1 E + a_2 = 0$$

From physical considerations, we know that the impulse response is a damped oscillation, so the roots E_1 and E_2 of the characteristic equation must be complex with magnitude less than one. This condition is equivalent to the condition that $a_2 < 1$ and $4a_2 - a_1^2 > 0$. The solution of the difference equation thus comes out to be

$$b(n) = e^{\lambda n} \frac{\sin(n+1)\omega_0}{\sin \omega_0}$$

where

$$\lambda = 0.5 \ln a_2$$

$$\omega_0 = \tan^{-1}\left[-a_1^{-1}\sqrt{4a_2 - a_1^2}\right]$$

The damped oscillation $b(n)$ is the impulse response function. The frequency ω_0 is the fundamental frequency of the impulse response function.

As we mentioned above, Yule ruled out superposed errors. Now, however, he allows a driving function (or input) of white noise, which he describes in the following way. The apparatus is left to itself, and unfortunately boys get into the room and start pelting the pendulum with peas, sometimes from one side and sometimes from the other. He states that the motion is now affected, not by superposed fluctuations, but by driving disturbances. As a result, the graph will be of an entirely different kind than a graph in the case of a sinusoid with superposed errors. The pendulum and pea graph will remain surprisingly smooth, but amplitude and phase will vary continuously, as governed by the inhomogeneous difference equation

$$x(n) + a_1 x(n-1) + a_2 x(n-2) = \epsilon(n)$$

where $\epsilon(n)$ is the white noise input. The solution of this difference equation is

$$x(n) = \sum_{k=0}^{\infty} b(k)\epsilon(n-k)$$

where $b(k)$ is the impulse response function given above.

Yule thus created a model with a finite number of parameters, namely, the coefficients a_1 and a_2 of the difference equation. Given an empirical time series $x(n)$, he uses the method of regression analysis to find these two coefficients. Because he regresses $x(n)$ on its own past instead of on other variables, it is a self-regression or autoregression. The least squares normal equations involve the empirical autocorrelation coefficients of the time series, and today these equations are called the Yule-Walker equations.

Yule carried out his autoregressive analysis on Wolfer's sunspot numbers, which are a sequence of yearly observations of sunspot observations. He used the numbers over the period 1749-1924 and obtained the autoregressive equation (with the mean value removed)

$$x(n) - 1.34254\, x(n-1) + 0.65504\, x(n-2) = \epsilon(n)$$

and thus

$$\lambda = 0.5 \ln 0.65504 = -0.21154$$

$$\omega_0 = 33.963° \text{ per year}$$

Hence, the dominant period is $360°/\omega_0 = 10.60$ years. Yule states that his autoregressive method represents an alternative method of estimating the spectrum, as opposed to the Schuster periodogram. In fact, his autoregressive model gives him an estimate not only of the power spectrum but of an amplitude-and-phase spectrum

$$B(\omega) = \sum_{n=0}^{\infty} b(n) e^{-i\omega n} = \frac{1}{1 + a_1 e^{-i\omega} + a_2 e^{-i\omega 2}}$$

which, for the sunspot numbers, is

$$B(\omega) = \frac{1}{1 - 1.34254\, e^{-i\omega} + 0.65504\, e^{-i\omega 2}}$$

The magnitude $|B(\omega)|$ and the phase $\theta(\omega)$ are given by the equation

Chapter 9. A Historical Perspective of Spectrum Estimation

$$B(\omega) = |B(\omega)| e^{i\theta(\omega)}$$

The power spectrum is the square of the magnitude spectrum, that is, $|B(\omega)|^2$. The peak is close to the fundamental frequency $\omega_0 = 33.963°$ per year. Except in exploration geophysics [19], [20], where Yule's amplitude-and-phase spectrum $B(\omega)$ is physically the spectrum of the minimum-delay seismic wavelet, Yule's spectral estimation method received scant attention until the 1960's.

Wiener's generalized harmonic analysis

Norbert Wiener [21] published in 1930 his classic paper, "Generalized Harmonic Analysis," which he personally considered his finest work. In his introduction, he states that he was motivated by the work of researchers in optics, especially that of Rayleigh and Schuster. However, Wiener demonstrated that the domain of generalized harmonic analysis was much broader than optics. Among Wiener's results was the writing down of the precise definitions of and the relationship between the autocovariance function and the power spectrum. The theorem that these two functions make up a Fourier transform pair is today known as the Wiener-Khintchine theorem [22].

Mention should be made of the basic fact that the existence of the spectrum follows from the properties of positive definite functions. Bochner's theorem on the spectral representation of positive definite functions provides a direct mathematical unification of spectral theories in Hilbert space and in stationary time series.

The writer several times in the 1950's discussed with Professor Wiener why his 1930 paper was not more accepted and used by the mathematical profession at the time. As with all things, Wiener looked at history quite objectively and with his characteristic concern for and love of people. In retrospect, it seems it was not until the publication of Wiener's book *Cybernetics* [23] in 1948 and also the non-classified publication of his book *Time Series* [24] in 1949 that the general scientific community was able to grasp the overall plan and implications of Wiener's contributions.

The following passage from Wiener's 1933 book *The Fourier Integral* [25] indicates the philosophy of Wiener's thinking and his great personal appeal:

"Physically speaking, this is the total energy of that portion of the oscillation lying within the interval in question. As this determines the energy-distribution of the spectrum, we may briefly call it the "spectrum." The author sees no compelling reason to avoid a physical terminology in pure mathematics when a mathematical concept corresponds closely to a concept already familiar in physics. When a new term is to be invented to describe an idea new to the pure mathematician, it is by all means better to avoid needless duplication, and to choose the designation already current. The "spectrum" of this book merely amounts to rendering precise the notion familiar to the physicist, and may as well be known by the same name."

Let us now define a stationary random process. We could either use discrete or continuous time, but for convenience let us use discrete (integer) time n. Let the process be denoted by $x(n)$, which, we will assume, has zero mean. The process is called (second-order) stationary, provided that its autocovariance function

$$\phi(k) = \mathbb{E}[x^*(n)\, x(n+k)]$$

depends only upon the time-shift k. Here, as always, the superscript asterisk indicates complex-conjugate. The normalized autocovariance function is called the autocorrelation function. However, Wiener generally used the term autocorrelation for $\phi(k)$, whether it was normalized or not. Nevertheless, it is confusing to keep using the term "autocorrelation" with two different meanings. It is better to use the term "autocovariance" wherever it is appropriate.

A white noise process is stationary. In the case of continuous time, its autocorrelation is $\delta(t)$ (the Dirac delta), whereas in the case of discrete time, its autocorrelation is δ_k (the Kronecker delta). As we have seen, the Fourier transform $E(\omega)$ of white noise in time is white in frequency; that is, the autocorrelation in frequency is the Dirac delta function:

$$\mathbb{E}[E^*(\omega)\, E(\omega + \mu)] = 2\pi \delta(\mu)$$

Chapter 9. A Historical Perspective of Spectrum Estimation

The problems confronting empirical workers in spectral analysis in the first part of the twentieth century were centered on the Schuster periodogram. Schuster introduced this concept at the turn of the century, and until Yule's work in 1927; it was the only method available to carry out empirical spectral analysis. Suppose that we observe a stationary random process for a very long time, so that we obtain a time series $x(n)$ for $n = 1, 2, \cdots, N$, where N is very large. Schuster then computed the periodogram

$$P(\omega) = \frac{1}{N} |X(\omega)|^2$$

where $X(\omega)$ is the discrete Fourier transform

$$X(\omega) = \sum_{n=1}^{N} x(n) e^{-i\omega n}$$

(Today we can compute $X(\omega)$ very rapidly by means of the Cooley-Tukey fast Fourier transform, but then it was a formidable task.) In the case when the stationary process is made up of sinusoidal waves with superimposed white noise, the periodogram is effective in picking out the discrete frequencies of the sinusoids. However, a purely nondeterministic stationary process is generated by the convolution formula (input-output relation)

$$x(n) = \sum_{k=1}^{N} b(k) \, \epsilon(n - k)$$

(Here we interpret $b(k)$ as the impulse response function of a filter, the white noise process $\epsilon(n)$ as the input to the filter, and the stationary process $x(n)$ as the output). For such a process, the Schuster periodogram $P(\omega)$ is extremely rough, and often cannot readily be interpreted. Empirical spectral analysis was at an impasse. Most of the time series observed in nature could not be analyzed by the methods available in 1930.

Now comes Wiener in 1930 with generalized harmonic analysis. In brief, Wiener in 1930 knew how to take the Fourier transform of a stationary random process, a milestone in the use of Fourier methods. Wiener's generalized harmonic analysis makes use of a generalized random function, namely, the Einstein-Wiener (white noise) process. In order to put Wiener's work into context, we will now give a small digression on the most widely known generalized function: the Dirac delta function.

The *impulse (Dirac delta) function* had been known for many years prior to its use by Dirac [26] in 1928. It was known by Heaviside [27]. However, it took the stature of a great physicist, Paul Dirac, to decree in 1928 the use of the impulse function in physics. In those early days, people used to talk about $\delta(t)$ as a function of t in the ordinary sense whose integral with $f(t)$ produces $f(0)$; that is,

$$\int_{-\infty}^{\infty} \delta(t) f(t) \, dt = f(0)$$

This idea used to cause great distress to mathematicians, some of whom even declared that Dirac was wrong despite the fact that he kept getting consistent and useful results. The physicists rejected these extreme criticisms and followed their intuition. We can now see why the physicists succeeded despite the reservations of the mathematicians. It is true that the physicists spoke of $\delta(t)$ as an ordinary function, which it cannot be in any precise sense, and that they treated it as an ordinary function by integrating it and even differentiating it. But the physicists were justified because they only used $\delta(t)$ inside integrals with sufficiently-differentiable functions $f(t)$. For example, the derivative of $\delta(t)$ always appeared inside an integral, and the integral was integrated by parts as follows:

$$\int_{-\infty}^{\infty} \delta'(t) f(t) \, dt = -\int_{-\infty}^{\infty} \delta(t) f'(t) \, dt = -f'(0)$$

The physicists never used the delta function except to map functions to real numbers. In this sense, they employed the machinery but not the words of distribution theory, which was devised expressly in order to give delta functions a sound basis. It was the French mathematician L. Schwartz who after World War II created a systematic theory of generalized functions and explained it in his well-known monograph *Theorie des Distributions* in 1950 and 1951. From then on the theory of generalized functions was developed intensively by many mathematicians. This precipitate development of distribution theory received its main stimulus from the requirements of mathematics and theoretical physics, in particular the theory of differential equations and quantum physics. Generalized functions possess a number of remarkable properties that extend the capabilities of classical mathematical analysis. For example, any generalized function turns out to be infinitely differentiable (in the generalized meaning); convergent series of generalized functions may be differentiated termwise an infinite number of times; the Fourier transform of a generalized function always exists; and so on. For this reason, the uses of generalized function techniques substantially expand the range of problems that can be tackled and leads to appreciable simplifications that make otherwise difficult operations automatic.

As science advances, its theoretical statements seem to require an ever higher level of mathematics. When he gave his theoretical prediction of the existence of antiparticles in 1931 (*Proc. Roy. Soc. London*, Ser. A, vol. 133, pp. 60-72) Dirac wrote, "It seems likely that this process of increasing abstraction will continue in the future and that advance in physics is to be associated with a continual modification and generalization of the axioms at the base of mathematics rather than with a logical development of any one mathematical scheme on a fixed foundation." Subsequent developments in theoretical physics have corroborated this view. In this essay, we have seen that since the time of Newton, the search for and the study of mathematical models of physical phenomena have made it necessary to resort to a wide range of mathematical tools and have thus stimulated the development of various areas of mathematics. Now let us return to Norbert Wiener in 1930.

Physicists are concerned with unlocking the mysteries of nature, and the impulse (Dirac delta) function eases their task. The impulse function is the simplest of the generalized functions. One can imagine the plight of Wiener in the mathematical community when he introduced generalized random functions into the mathematical literature as early as 1923, and especially in his 1930 paper.

Let us now give the gist of Wiener's generalized harmonic analysis. As we know, convolution in the time domain corresponds to multiplication in the frequency domain. Thus in terms of Fourier transforms, the above input-output convolution integral becomes

$$X(\omega) = B(\omega) E(\omega)$$

In this equation, $E(\omega)$ is the Fourier transform of white noise, so that $E(\omega)$ is a generalized function that is white (i.e., very rough) in frequency. The filter's transfer function $B(\omega)$ is a smooth well-behaved (ordinary) function. The product $X(\omega)$ is also very rough.

Let us now take the inverse Fourier transform of $X(\omega)$. It is

$$x(n) = \frac{1}{2\pi} \int_{-\pi}^{\pi} e^{i\omega n} X(\omega) d\omega$$

which is

$$x(n) = \frac{1}{2\pi} \int_{-\pi}^{\pi} e^{i\omega n} B(\omega) E(\omega) d\omega$$

This formula represents Wiener's generalized harmonic analysis of $x(n)$; that is, it is the spectral representation of the stationary random process $x(n)$. It involves the smooth filter transfer function $B(\omega)$ and the very rough (white in frequency) process $E(\omega)$. We thus see that the spectral representation requires Wiener's generalized random function $E(\omega)$, which came out of his studies of Brownian movement.

Wiener's generalized harmonic analysis (i.e., spectral representation) explains why the periodogram of Schuster did not work for convolutional processes. Because the periodogram (as the number of observations becomes large) is

$$P(\omega) = \frac{1}{N} |X(\omega)|^2$$

it follows that the periodogram has the intrinsic roughness of the $X(\omega)$ process. (It was not until the work of J. Tukey [34] in 1949 that a means was found to overcome this problem; Tukey's breakthrough was of epoch proportions.)

Wiener in his 1930 paper gave the following method, which was standard until the work of Tukey in 1949. Wiener's method was intended for very long time series. It consisted of computing the autocovariance function as the time average

Chapter 9. A Historical Perspective of Spectrum Estimation

$$\phi(k) = \frac{1}{N}\sum_n x^*(n)x(n+k)$$

for $-p \leq k \leq p$, where p is less than the data length N, and then computing the power spectrum $\Phi(\omega)$ as the Fourier transform

$$\Phi(\omega) = \sum_{k=-p}^{p} \phi(k)e^{-i\omega k}$$

This Fourier transform relationship between autocovariance and power spectrum, as we have observed, is now called the Wiener-Khinchin theorem.

Whereas von Neumann's work in quantum physics in 1929 received 'instant acclaim and well-deserved recognition by physicists and mathematicians, Wiener's work in 1930 lay dormant. However, now with the benefit of hindsight, it is worthwhile for us to reconcile these two approaches to spectral estimation. This we will do in the next section.

Reconstruction of the two spectral theories

We have come a long way in the history of spectral estimation to this point. From the work of the ancients in deriving a calendar, to the work of the great mathematicians who formulated the wave equation in the eighteenth century, it took thousands of years. Then the work of Bernoulli, Euler, and Fourier came, and the result was a spectral theory in terms of sinusoidal functions, in place at the beginning of the nineteenth century. The theory was extended to the case of arbitrary orthogonal functions by Sturm and Liouville, and this led to the greatest empirical success of spectral analysis yet obtained: the physical results of spectral estimation that unlocked the secret of the atom. Credit for this result belongs to Heisenberg and Schrodinger in 1925 and 1926. Then in 1929, the work of von Neumann put the spectral theory of the atom on a firm mathematical foundation in his spectral representation theorem. The spectral work of von Neumann represents the cumulation of this line of research in quantum physics. Meanwhile, Rayleigh and Schuster at the beginning of the twentieth century were applying the original sinusoidal methods of Fourier to the analysis of data in the realm of classical physics. However, the periodogram approach of Schuster did not work well for purely nondeterministic stationary random processes, and this led Yule in 1927 to develop a spectral theory for a subclass known as autoregressive processes. Meanwhile, Wiener had developed the mathematical theory of Brownian movement in 1923, and in 1930 he introduced generalized harmonic analysis; that is, the spectral representation of a stationary random process. Thus in 1930, we have two spectral theories, one represented by the spectral representation theorem of von Neumann and the other by the spectral representation theorem of Wiener. It is the purpose of this section, with the benefit of hindsight, of course, to indicate the relationship between von Neumann and Wiener spectral theories.

The common ground is the Hilbert space. As we have seen, the von Neumann result is the spectral representation of a Hermitian operator H in Hilbert space. The Schrodinger equation is written in terms of a Hermitian operator, and this equation governs the spectrum of atoms and molecules. Now let us, however, leave this Hilbert space and look at another one. The other Hilbert space is one defined by the probability measure that governs the stationary random process in question. As we know, a Hilbert space is specified by an inner (or dot) product. The elements of the Hilbert space are random variables, and the inner product is defined as the expected value given by

$$\langle x, y \rangle = \mathbb{E}\, x^* y$$

(The superscript asterisk indicates the complex conjugate.) In this Hilbert space, a stationary process is defined as follows. We use discrete (integer) time n, although a similar development can be made in the case of continuous time. A sequence of random variables $x(n)$ in Hilbert space is called a *stationary random process* if its autocovariance

$$\phi(k) = \langle x(n), x(n+k) \rangle$$

depends only upon the time-shift k and not on absolute time n. This definition implies that the elements $x(n)$ of the process are generated recursively by a unitary operator; that is,

$$U x(n) = x(n+1)$$

so that

Chapter 9. A Historical Perspective of Spectrum Estimation

$$x(n+k) = U^k x(n)$$

Because a unitary operator represents a rotation, we see that a stationary random process traces out a spiral in Hilbert space, the so-called Wiener spiral. We now come to the connection that we are seeking, namely, the fact that the Cayley-Mobius transformation [28] of a Hermitian operator is a unitary operator. Thus there is a one-to-one correspondence between Hermitian operators and unitary operators in Hilbert space. The von Neumann spectral representation is for a Hermitian operator. If we take its Cayley-Mobius transformation, we obtain the corresponding spectral representation for the unitary operator U. This spectral representation has the form

$$U = \frac{1}{2\pi} \int_{-\pi}^{\pi} e^{i\omega} \mathbb{U}(\omega) d\omega$$

where $\mathbb{U}(\omega)$ represents a family of projection operators as a function of circular frequency ω. Thus the process has the representation

$$x(n) = U^n x(0) = \frac{1}{2\pi} \int_{-\pi}^{\pi} e^{i\omega n} \mathbb{U}(\omega) x(0) d\omega$$

We now make the identification

$$\mathbb{U}(\omega) x(0) = X(\omega)$$

and we obtain

$$x(n) = \frac{1}{2\pi} \int_{-\pi}^{\pi} e^{i\omega n} X(\omega) d\omega$$

This equation is Wiener's generalized harmonic analysis of the process. Thus we have the connection we sought; the two spectral representations are related by the Cayley-Mobius transformation.

Wiener-Levinson prediction theory

Early in 1940, Wiener became involved in defense work at MIT and, in particular, he became interested in the design of fire-control apparatus for anti-aircraft guns. The problem was to build into the control system of the gun some mechanical device to aim the gun automatically. The problem, in effect, was made up of two parts: a mathematical part, which consisted of predicting the future position of an airplane from its observed past positions, and an engineering part, which consisted of realizing the mathematical solution in the form of an actual physical device. Wiener recognized that it was not possible to develop a perfect universal predictor, and so he formulated the mathematical problem on a statistical basis. He defined the optimum predictor as the one that minimizes the mean-square prediction error. The minimization led to the Wiener-Hopf integral equation, which represented the completion of the mathematical part of the problem. As to the engineering part, Wiener immediately recognized that it was possible to devise a hardware apparatus that represents the solution to the Wiener-Hopf equation. As Wiener [29] states in his autobiography (p. 245): "It was not hard to devise apparatus to realize in the metal what we had figured out on paper. All that we had to do was make a quite simple assembly of electric inductances, voltage resistances, and capacitors, acting on a small electric motor of the sort which you can buy from any instrument company." Wiener's mathematical results [24] were published in 1942 as a classified report to Section D2 of the National Defense Research Committee. This report is Wiener's famous Time Series book, which we mentioned previously. Its full title is *Extrapolation, Interpolation, and Smoothing of Stationary Time Series with Engineering Applications*, and it was republished as an unrestricted document in 1949 by MIT Press, Cambridge.

Although Wiener's "General Harmonic Analysis" did not have immediate influence, his *Time Series* book, which was written in a more understandable style, did among those who had access to the book in 1942 and the general public in 1949. As we will now see, a great deal of credit for the dissemination of Wiener's ideas belongs to his former student and his colleague, Professor N. Levinson.

Levinson's initial contact with Wiener was in Wiener's course in 1933-34 on Fourier Series and Integrals, which is described in Levinson's own words as follows:

> "I became acquainted with Wiener in September 1933 while still an undergraduate student of electrical engineering, when I enrolled in

his graduate course. It was at that time really a seminar course. At that level he was a most stimulating teacher. He would actually carry out his research at the blackboard. As soon as I displayed a slight comprehension of what he was doing, he handed me the manuscript of Paley-Wiener for revision. I found a gap in a proof and proved a lemma to set it right. Wiener thereupon set down at his typewriter, typed my lemma, affixed my name and sent it off to a journal. A prominent professor does not often act as a secretary for a young student."

N. Levinson, a dynamic and brilliant mathematician and a warm and kind person, made important and permanent contributions to engineering and applied science. The Levinson theorem in quantum mechanics illustrates his ability to grasp the relationship between physical concepts and mathematical structure. Few have this insight, and nowhere is it better demonstrated than in the two expository papers written in 1942 by Levinson soon after the restricted publication of Wiener's *Time Series* book. These two papers were published in 1947 in the *Journal of Mathematical Physics*, and thus they represented the first public disclosure of Wiener's time series results. Later these two papers also appeared as Appendices C and B in the unrestricted publication of Wiener's book in 1949 by MIT Press [24].

An appreciation of Levinson's contribution can be gained in historical perspective. The 1942 edition of Wiener's book was bound in a yellow paper cover, and because of its difficult mathematics, it came to be known among engineers as the "yellow peril" (a term familiar to mathematicians as applying to a famous series of advanced texts). However clear in a conceptual way the building of an actual device was to Wiener, there were few engineers at that time who were able to grasp Wiener's mathematical solution, much less to realize it in the form of a physical device. At this point, Levinson stepped in and wrote "A heuristic exposition of Wiener's mathematical theory of prediction and filtering," [30], one of his two classic applied papers on explaining Wiener's work. Levinson describes his paper as an expository account of Wiener's theory. Levinson's earlier training was in electrical engineering, so he understood hardware design methods. In the paper, Levinson shows in an elementary way why the Wiener-Hopf equation cannot be solved by use of the Fourier transform theorem. Then, in a natural way, he introduces the spectral factorization and obtains the explicit solution for the prediction operator and, more generally, for the filter operator. This masterpiece of exposition opened up these methods to the engineering profession.

Levinson's other classic applied paper is entitled, "The Wiener RMS (root mean square) error criterion in filter design and prediction." [31] As before, let us try to put this paper in historical perspective. In 1942, the Army Air Force Weather Division negotiated a contract with MIT to perform statistical analyses of meteorological and climatological data, particularly in relationship to weather forecasting, and to conduct research into the application of statistical techniques to long-range forecasting [32]. Professor G. P. Wadsworth of the MIT Mathematics Department was in charge of this meteorological project. The basic idea was to collect and sort large amounts of numerical meteorological data and to forecast by analogy, much like the forecasts made on television today, in which the weatherman looks at the data appearing on the satellite picture of the earth. Wadsworth's method had merit, but the data required was just not available in the 1940's. Wiener's *Time Series* book was completed at about the same time as this MIT Meteorological Project was starting up. Since the weather data available occurred at discrete intervals of time, the continuous-time methods of Wiener were not directly applicable. As a result, Wadsworth asked Levinson to write up a discrete form of Wiener's theory. The result was Levinson's "Wiener RMS" paper with the Levinson recursion. However, use was never made of Wiener-Levinson prediction theory by the MIT Meteorological Project, and Levinson's paper sat dormant.

In order to understand why Levinson's methods were not used in the 1940's, one must look at the computing facilities available at the time. The actual realization of these methods would have to be carried out by people using hand calculators. A hand calculator could add, subtract, multiply, and divide, but had no memory except an accumulator. Thus the result of each separate calculation had to be transferred to paper by hand, a drawn-out, time-consuming process. In contrast to the hardware devices working in real time, as envisaged by Wiener, the

Chapter 9. A Historical Perspective of Spectrum Estimation

hand calculator was a poor substitute. As Wiener [24, p. 102] states: "Much less important, though of real interest, is the problem of the numerical filter for statistical work, as contrasted with the filter as a physically active piece of engineering apparatus."

After Levinson wrote his two expository papers, which were completed in 1942, neither he nor Wiener took up research in the computational (software) aspect of Wiener's theory. Wiener was more interested in its realization by machines (hardware), and his research interest was already shifting to biological and medical problems. In fact, it was the union of these two research interests that led to his discovery and formulation of the science of cybernetics, which he describes as the problem of control and communication in machines and animals. Meanwhile Levinson had decided as early as 1940 to shift his field from the Fourier methods of Wiener to the field of nonlinear differential equations. He talked about this decision with his friends in 1940. Levinson worked hard over a period of two or three years (which included the period during which he wrote the two expository papers) before he felt that he had enough mastery in his new field. Such mastery he did achieve and his outstanding contributions to differential equations were recognized by his receiving the prestigious Bocher Prize in Mathematics in 1954.

Despite their other research interests, both Wiener and Levinson were always ready to give their support and time to the MIT Meteorological Project directed by G. P. Wadsworth. Wiener was especially interested in seeing physical examples of autocorrelation functions. This interest led to the computation of several autocorrelation functions of ocean wave data by Wadsworth and by his friend and associate H. R. Seiwell [33], who was with the Woods Hole Oceanographic Institution. The interest in these computations led to the "Symposium on Autocorrelation Analysis Applied to Physical Problems" held at Woods Hole, MA in June 1949, sponsored by the Office of Naval Research. The high point of this meeting was the paper by Tukey [34]. Before Tukey's work, the power spectra computed from empirical autocorrelation functions were too erratic to be of any use in formulating physical hypotheses. Not only did Tukey show correctly how to compute power spectra from empirical data, but he also laid the statistical framework for the analysis of short-time series, as opposed to the very long ones envisaged by Wiener and Levinson.

Wadsworth was also director of the MIT section of the U.S. Naval Operations Evaluation Group, a project started in World War II which initiated the use of operations research in the United States. By 1950, Wadsworth was applying operations research methods to industry and had established himself as one of the highest paid consultants in the United States. There were so many industrial people waiting to see him in his outer office at MIT that one had to make an appointment with his secretary many weeks in advance to see him in his inner office for just 5 or 10 minutes at the most. The writer began as one of Wadsworth's research assistants in the MIT Mathematics Department in September 1950, and he was assigned to work in seismology by Professor Wadsworth. Magnolia Petroleum made available eight seismic records, and the writer immediately got a very lonely feeling, especially at MIT at night digitizing the Magnolia seismic records with a ruler and pencil. Except for Wadsworth, in 1950 nobody at MIT or in the oil industry thought that the analysis of digital seismic data would ever be feasible.

Fortunately Tukey took an interest in the seismic project and conveyed his research ideas by mail. The first empirical results were the computation of the Tukey spectra for various sections of the Magnolia records in the spring of 1951. From these spectral results, a seismic analysis based on prediction error was formulated in the summer of 1951. This analysis made use of Wiener prediction theory in digital form. Prior to this work, Wiener's procedures had only been realized in analog form. In hand plotting the first numerical results of what today is called predictive deconvolution, or blind deconvolution, or linear predictive coding (LPC), the writer was so amazed that he could not believe his eyes, and he was sure that he would never see such good results again. But the second trace, and the third trace, and so on, were computed and confirmed what he saw. The digital signal processing method of deconvolution worked! As soon as possible, he made an appointment to see Wadsworth, which the secretary set three weeks from then, in September 1951. The result was that the digitally-processed seismic

Chapter 9. A Historical Perspective of Spectrum Estimation

traces [35] were sent out to the oil industry, and the oil companies gave money to support a project. The MIT Geophysical Analysis Group was thus born, and the MIT Whirlwind digital computer was continuously used by the GAG to digitally process seismic records and to do other geophysical research. During this period, Tukey freely gave his research advice [36], [37]. For example, Tukey's methods for estimating coherency (today called by various names by the oil industry) are vital in the estimation of seismic velocity as well as in other multichannel methods. Tukey's vision of a fast Fourier transform was always influential. In fact, S. M. Simpson [38], who later directed the Geophysical Analysis Group, eventually devised an efficient 24-point Fourier transform, which was a precursor to the Cooley-Tukey fast Fourier transform in 1965. The FFT made all of Simpson's efficient autocorrelation and spectrum programs instantly obsolete, on which he had worked the equivalent of half a lifetime. Wiener was very generous of his time [39]. Wiener's work [40], [41] on multichannel methods was helpful later in extending the Levinson recursion, which Levinson had devised for single channel time series, to the multichannel case [42]. The excellent seismic data and corresponding well logs supplied by the oil industry to the Geophysical Analysis Group made possible the development of the statistical minimum-delay model of the earth's strati-graphic layers, together with the theoretical justification of seismic deconvolution [19], [43].

In the late 1950's a digital revolution occurred because of the introduction of transistors in the building of digital computers, which made possible reliable computers at a much lower cost than previously. As a result, the seismic industry completely converted to digital technology in the early 1960's, a long ten years after the first digital results were obtained. Since then, nearly every seismic record taken in the exploration of oil and natural gas has been digitally deconvolved and otherwise digitally processed by these methods. The final result of the digital processing of seismic data was the discovery of great oil fields which could not be found by analog methods. These oil fields include most of the offshore discoveries, as in the North Sea, the Gulf of Mexico, the Persian Gulf, as well as great onshore discoveries in Alaska,

Asia, Africa, Latin America, and the Middle East, made since the digital revolution that took place in geophysical exploration in the early and mid 1960s . Today [which refers to the year 1982 when this paper was written] an oil company will deconvolve and process as many as a million seismic traces per day; it took a whole summer in 1951 to do 32 traces.

Whereas the digital revolution came first to the geophysical industry largely because of the tremendous accuracy and flexibility afforded by large digital computers, today we are in the midst of a universal digital revolution of epic proportions. One now realizes that the work of Wiener and Levinson is being appreciated and used by an ever-increasing number of people. Digital signal processing is a growing and dynamic field which involves the exploration of new technology and the application of the techniques to new fields. The technology has advanced from discrete semiconductor components to very large-scale integration (VLSI) with densities above 100 000 components per silicon chip. The availability of fast, low-cost microprocessors and custom high-density integrated circuits means that increasingly difficult and complex mathematical methods can be reduced to hardware devices as originally envisaged by Wiener, except the devices are digital instead of analog. For example, a custom VLSI implementation of linear predictive coding is now possible, requiring a small number of custom chips. Whereas originally digital methods were used at great expense only because the application demanded high flexibility and accuracy, we have now reached the point that anticipated long-term cost advantages have become a significant factor for the use of digital rather than analog methods.

Tukey empirical spectral analysis

As we have mentioned, a turning point in the empirical analysis of time series data began in 1949 at the Woods Hole Symposium on Applications of Autocorrelation Analysis. There Tukey presented the first of three papers [34], [36], [37], which he had written in the early years on spectrum analysis. These papers introduced the classic Tukey method of numerical spectral estimation, a method that has been used

Chapter 9. A Historical Perspective of Spectrum Estimation

by most workers since that time. In addition, Tukey described an approximate distribution for the estimate. This distribution was required for the proper design of experiments for the collection of time series data. In a very interesting paper [44], Tukey describes the situation which led to his spectral work, including a discussion of Hamming's suggestion about the smoothing of the discrete Fourier transform of an empirical autocorrelation, which led to the joint work of Hamming and Tukey.

During the last four decades, Tukey [45]—[50] introduced a multitude of terms and techniques that are standard to the practice of the data analysis of time series. Such commonplace terms and concepts as "prewhitening," "aliasing," "smoothing and decimation," "tapering," "bispectrum," "complex demodulation," and "cepstrum" are due to Tukey. Very few papers in the literature of applied time series analysis do not give some acknowledgment of Tukey's ideas and methods, and most papers credit his ideas in some vital way. Moreover, Tukey [51]-[55] has made substantial contributions in the placing of the data analysis of time series into perspective with current research in the physical sciences, in statistics, and in computing and numerical analysis.

We have already mentioned the key influence Tukey had on the MIT group, which included Wadsworth, Simpson, the writer, and others. W. J. Pierson and L. J. Tick [56] at New York University used Tukey's methods in the analysis of oceanographic time series records. The outstanding thesis of Goodman [57], which extended the results of Tukey to the bivariate case, was written under Tukey's supervision. The group at La Jolla, CA, which included Munk, Rudnick, and Snodgrass, applied Tukey's spectral methods to estimate wave motion due to storms many thousands of miles away, a testimony to the power of his methods. Munk and McDonald wrote a remarkable book, *The Rotation of the Earth* [58], which used these spectral methods in several novel ways.

In this period, packages of computer programs for time series analysis were appearing. The collection of programs by Healy [59] of Bell Laboratories were circulated from 1960 on. The BOMM collection of programs [60] was developed at La Jolla. Some of the programs used by

Parzen were included in his book [61]. The programs written at MIT are described in [38] and [62].

In econometrics, Granger's book [63] in 1964 described many of the techniques suggested by Tukey for the analysis of univariate and bivariate time series. The most successful application of spectrum techniques to economic series is its use for the description of the multitude of procedures of seasonal adjustment. In astronomy, Neyman and Scott [64] in 1958 carried out the analysis of two-dimensional data consisting of the positions of the images of galaxies on photographic plates.

Norbert Wiener remained active until his death in 1964. His later work included both empirical results, such as modeling and analyzing brain waves [66], [67], and theoretical results, such as his work with Masani on multivariate prediction theory [40], [41]. Wiener's death marks the end of an era in time series analysis and spectral theory.

Cooley-Tukey fast Fourier transform

The present epoch of time series analysis began in 1965 with the publication of the fast Fourier transform by Cooley and Tukey [68]. The effect that this paper has had on scientific and engineering practice cannot be overstated. The paper described an algorithm for the discrete Fourier transform of $T = T_1 \cdots T_p$ values by means of $T(T_1 + \cdots + T_p)$ multiplications instead of the naive number T^2. Although such algorithms existed previously [69], they seem not to have been put to much use. Sande developed a distinct, symmetrically related algorithm simultaneously and independently.

The existence of such an algorithm meant, for example, that the following things could be computed an order of magnitude more rapidly: spectrum estimates, correlograms, filtered versions of series, complex demodulates, and Laplace transforms (see, for example, [70]). General discussions of the uses and importance of fast Fourier transform algorithms may be found in [71] and [72]. The Fourier transform of an observed stretch of series can now be taken as a basic statistic and classical statistical analyses—such as multiple regression, analysis of variance, principal components, canonical analysis, errors in

Chapter 9. A Historical Perspective of Spectrum Estimation

variables, and discrimination—can be meaningfully applied to its values, [73] and references cited therein. Higher-order spectra may be computed practically [74]. Inexpensive portable computers for carrying out spectral analysis have appeared on the market and may be found in many small laboratories.

The years since 1965 have been characterized by the knowledge that there are fast Fourier transform algorithms. They have also been characterized by the rapid spread of type of data analyzed. Previously, the data analyzed consisted almost totally of discrete or continuous real-valued time series. Now the joint analysis of many series, such as the 625 recorded by the Large Aperture Seismic Array in Montana [75], has become common. Spatial series are analyzed [75]. The statistical analysis of point processes has grown into an entirely separate field [76]. The SASE IV computer program developed by Peter Lewis [77] has furthered such analysis considerably. We note that transforms other than the Fourier are finding interest as well [78], [79].

Predictive deconvolution (aka blind deconvolution)

In this section we want to give the Robinson method of predictive deconvolution [19] [80] (aka predictive decomposition). The seismic trace, after signature deconvolution, has the form

$$x(k) = b(k) * \epsilon(k) \qquad (1)$$

In the general case we do not have an estimate of the reverberation waveform $b(k)$, nor do we know the reflectivity function $\epsilon(k)$. Given the signal $x(k)$, we want to unravel it so as to obtain $x(k)$ and $\epsilon(k)$. How do we do this? The method of predictive deconvolution makes use of two hypotheses, namely, the feedback hypothesis and the random hypothesis. Specifically these two hypotheses are respectively:

(1) The reverberation wavelet is the impulse response of a finite feedback system, and thus is minimum delay (aka minimum phase).

(2) The reflectivity function, at least within certain time gates, has the properties of a white-noise random process.

On the basis of these two hypotheses, we will see how to find the deconvolution operator a_k which yields the reflectivity function from the seismic trace; i.e.

$$a_k * x(k) = \epsilon(k) \qquad (2)$$

The sedimentary layers in producing a reflection seismic trace act as a time-varying filter. However, if a given time gate has a reflectivity function made up of only small reflection coefficients, then the time-varying filter reduces approximately to a time-invariant finite feedback filter. If, in addition, the reflectivity-function within the given time gate is a white-noise random process, then both hypotheses required for predictive deconvolution are upheld. Fortunately geophysical experience has shown that such time gates can usually be found on nearly all seismic traces taken in petroleum exploration, and as a result predictive deconvolution is routinely carried out on all seismic records in every exploration program.

In mathematical terms, the hypothesis that the filter is a finite feedback (AR filter) means that its transfer function has the form

$$B(z) = \frac{1}{1 + a_1 z^{-1} + a_2 z^{-2} + \cdots + a_p z^{-p}} \qquad (3)$$

where the feedback parameters $a_1, a_2, \cdots, a_p$ are constants. The order p of the system is either fixed in advance, or determined by a mean-square error criterion. Given the seismic trace $x(k)$, the problem is to determine the values of the feedback parameters. These parameters are the coefficients of the inverse (i.e. deconvolution) filter, i.e.

$$A(z) = 1 + a_1 z^{-1} + a_2 z^{-2} + \cdots + a_p z^{-p} \qquad (4)$$

Once we calculate these parameters, the *deconvolution process* is then carried out, i.e.

$$\sum_{n=0}^{p} a_k u(k-n) = \epsilon(k) \quad \text{(where } a_0 = 1\text{)} \qquad (5)$$

Thus the result of the deconvolution is the reflectivity function $\epsilon(k)$.

How do we find the values of the feedback parameters? This is the question we want to answer in this section.

Chapter 9. A Historical Perspective of Spectrum Estimation

Given that the seismic trace (after signature deconvolution) is

$$x(k) = b(k) * \epsilon(k)$$

we know that the autocorrelations satisfy

$$\phi_{xx}(k) = \phi_{bb}(k) * \phi_{\epsilon\epsilon}(k)$$

and the z-transforms satisfy

$$\Phi_{xx}(z) = \Phi_{bb}(z)\, \Phi_{\epsilon\epsilon}(z)$$

The feedback hypothesis means that

$$\Phi_{bb}(z) = B(z)\, B(z^{-1}) = \frac{1}{A(z)A(z^{-1})}$$

whereas the random hypothesis means that

$$\Phi_{\epsilon\epsilon}(z) = \sigma^2 = \text{constant}$$

Thus we have

$$\Phi_{xx}(z) = \frac{\sigma^2}{A(z)A(z^{-1})}$$

The problem now is to solve this equation for σ^2 and $A(z)$. For simplicity we will drop the subscripts on $\Phi_{xx}(z)$ and denote it by $\Phi(z)$. Likewise we drop the subscripts on $\phi_{xx}(k)$ and denote it by $\phi(k)$. Thus the above equation is

$$\sum_{n=-\infty}^{\infty} \phi(n)\, z^{-n} = \frac{\sigma^2}{A(z)A(z^{-1})}$$

We now multiply both sides of this equation by $A(z)$ and obtain

$$A(z) \sum_{n=-\infty}^{\infty} \phi(n)\, z^{-n} = \frac{\sigma^2}{A(z^{-1})} \tag{6}$$

Because

$$A(z) = 1 + a_1 z^{-1} + a_2 z^{-2} + \cdots + a_p z^{-p}$$

is a minimum-delay polynomial in the variable z^{-1} [81], it follows that

$$B(z) = \frac{1}{A(z)} = 1 + b(1) z^{-1} + b(2) z^{-2} + b(3) z^{-3} + \cdots$$

is a power series in the same variable. Replacing z^{-1} by z in the above equation, we obtain

$$\frac{1}{A(z^{-1})} = 1 + b(1)z + b(1)z^2 + b(1)z^3 + \cdots = \sum_{k=-\infty}^{0} b(-k)\, z^{-k}$$

which is a power series in z. Note that $f(0) = 1$. Thus the right-hand side of equation (6) is

$$\sum_{k=-\infty}^{0} [\sigma^2 b(-k)]\, z^{-k} \qquad (7)$$

Now let us look at the left-hand side of equation (6.6). As we know, multiplication in the z-domain corresponds to convolution in the time domain. Thus the multiplication of $A(z)$ with $\Phi(z)$ corresponds to the convolution of their coefficients. If we denote their product by $\Psi(z)$, we can write

$$A(z)\Phi(z) = \Psi(z)$$

which more explicitly is

$$\left[\sum_{n=0}^{p} a_k\, z^{-k}\right]\left[\sum_{k=-\infty}^{\infty} \phi(k)\, z^{-n}\right] = \sum_{k=-\infty}^{\infty} \psi(k)\, z^{-k}$$

Note that $a_0 = 1$. Thus $\psi(k)$ is given by the convolution

$$\psi(k) = a_k * \phi(k)$$

which is

$$\psi(k) = \sum_{n=0}^{p} a_n\, \phi(k - n)$$

Thus the left-hand side of equation (6) is

$$\sum_{k=-\infty}^{\infty}\left[\sum_{n=0}^{p} a_n\, \phi(k-n)\right] z^{-k} \qquad (8)$$

Combining equations (7) and (8) we have

$$\sum_{k=-\infty}^{\infty}\left[\sum_{n=0}^{p} a_n\, \phi(k-n)\right] z^{-k} = \sum_{k=-\infty}^{0} [\sigma^2 b(-k)]\, z^{-k}$$

Chapter 9. A Historical Perspective of Spectrum Estimation

Because this equation is an identity in the variable z^{-1}, the coeficient on the left must equal the coefficient on the right for each value of k. We thus have

$$\sum_{n=0}^{p} a_n \phi(k-n) = \sigma^2 b(-k) \quad \text{for} \quad k = 0, -1, -2, \cdots \quad (9)$$

and

$$\sum_{n=0}^{p} a_n \phi(k-n) = 0 \quad \text{for} \quad k = 0, 1, 2, \cdots \quad (10)$$

This set represents an infinite set of equations. The usual situation is the following. We compute the autocorrelation values

$$\phi(0), \phi(1), \phi(2), \cdots, \phi(p)$$

from the seismic trace $x(k)$. By symmetry we also know the values of

$$\phi(-1), \phi(-2), \cdots, \phi(-p)$$

We then make use of a subset of the above infinite set of equations. This subset is made up of the equations for $k = 0, 1, 2, \cdots, p$, and is

$$\phi(0) + a_1\phi(-1) + a_2\phi(-2) + \cdots + a_p \phi(-p) = \sigma^2$$
$$\phi(1) + a_1\phi(0) + a_2\phi(-1) + \cdots + a_p \phi(-p+1) = 0$$
$$\phi(2) + a_1\phi(1) + a_2\phi(0) + \cdots + a_p \phi(-p+2) = 0$$
$$\cdots \quad \cdots$$
$$\phi(p) + a_1\phi(p-1) + a_2\phi(p-2) + \cdots + a_p \phi(0) = 0$$

These simultaneous equations are known as the *normal equations*. We can solve these equations for the unknown σ^2 and unknown $a_1, a_2, \cdots, a_p$. Thus we have found the required coefficients of the deconvolution operator. The use of this operator on $x(k)$ yields the reflectivity function, as seen by equation (5).

An interesting point now comes up. What do the rest of the equations in set (10) mean? The remaining equations (i.e. the equations (10) for $k = p+1, p+2, \cdots$) are

$$\phi(p+1) + a_1\phi(p) + a_2\phi(p-1) + \cdots + a_p \phi(1) = 0$$
$$\phi(p+2) + a_1\phi(p+1) + a_2\phi(p) + \cdots + a_p \phi(2) = 0$$

... ...

The first of these equations can be used to find $\phi(p+1)$ and the next equation can be used to solve for $\phi(p+2)$. Using these equations in turn, we can thus find all the remaining values of the autocorrelation function. The autocorrelation function found in this way is known as the maximum entropy autocorrelation. Its Fourier transform is the maximum entropy spectrum.

The following critique is given by Schneider [82]:

> "The work horse of statistical wavelet deconvolution for the past one and one half decades has been the predictive decomposition approach, which assumes the reflectivity function is statistically white and the convolutional wavelet to be minimum-phase. To say that this has not been an effective tool is to condemn hundreds of thousands of miles of seismic processing and to deny untold millions of barrels of oil discovered from these data."

Silvia and Robinson [83], through the use of lattice methods, have related the concept of the autoregressive analysis to the geophysical inverse problem. Itakura and Saito [84] were responsible for introducing two important ideas into spectrum estimation that are now gaining wide acceptance in the engineering world. The first idea is that of using maximum likelihood in spectrum estimation. Although the idea itself was not new, their introduction of a particular spectrum distance measure is becoming more and more important for different applications, such as speech. Parzen [85] gives the name information divergence to this measure, which is the same as the Kullback-Leibler information number. He also shows its relation to the notion of cross-entropy. The second idea is that of using the lattice as a filter structure for the purpose of analysis (as an all-zero filter) and synthesis (as an all-pole filter). The idea of an adaptive lattice was first proposed by Itakura and Saito as a way of estimating the partial correlation (PARCOR) coefficients (a term they coined) adaptively. Makhoul [86] gives stable and efficient lattice methods. The lattice has become important because of its fast convergence and its relative insensitivity to round-off errors.

Statistical theory of spectrum estimation

Since the pioneering work of Tukey [34] in 1949, many important contributions have been made to the statistical theory of spectrum estimation. An adequate treatment would require a long paper in itself, and so all we can hope to do here is to raise the reader's consciousness concerning the statistical theory required to understand and implement spectrum estimation.

The writer has great admiration for the work of Parzen, who from the 1950's to the present time has consistently made bedrock contributions both in theory and applications [61], [85], [87], [88]. His long series of papers on time series analysis include the famous Parzen window for spectrum analysis. Another one of Parzen's important contributions is his formulation of the time series analysis problem in terms of reproducing kernel Hilbert spaces. A remarkable number of Ph.D. theses on time series analysis have been written under the direction of Parzen, more than any other person. The writer has had the good fortune to discuss geophysical time-series problems with Professor Parzen over the years, and in every case Parzen has been able to provide important physical insight in the application of the statistical methods. The Harvard lectures by Professor Parzen in 1976 represent one of the high points in time series analysis and spectrum estimation ever to be heard in those venerable halls.

The book by Grenander and Rosenblatt [65] in 1957 formalized many of the data analysis procedures and approximations that have come into use. They have an extensive treatment of the problem of choice of window and bandwidth. The further contributions to this problem by Parzen and by Jenkins are discussed in the 1961 paper of Tukey [51]. An accurate and informative account of the developments of spectrum estimation in the 1950's is given by Tukey [44].

Another important statistical development that deserves mention is the alignment issue in the estimation of coherence, worked on in the 1960's by Akaike and Yamanouch, by Priestley, and by Parzen. Discussions of this work and the references can be found in the book by Priestley [89]. This excellent book, which appeared in 1981, has already set a new standard. It can be recommended as an authoritative account of the statistical theory of spectrum estimation, which we only touch upon in this section.

H. Wold coined the names "moving average process" and "autoregressive process" in his 1938 thesis [90] under Professor Harald Cramer at Stockholm University. In his thesis, Wold computed a model of the yearly level of Lake Vaner in Sweden as a moving average of the current rainfall and the previous year's rainfall. He also computed an autoregressive model of the business cycle in Sweden for the years 1843-1913. In turn, Whittle wrote his 1951 thesis [91] under Professor Wold at Uppsala University. Whittle opened up and made important contributions to the field of hypothesis testing in time-series analysis. Whittle's careful work is exemplified by his autoregressive analysis of a seiche record [92] in which he fits a low level autoregressive model to the data and gives statistical tests to determine the appropriateness of the model. Professor Whittle used to return to Sweden for visits, and the writer remembers taking long walks with him through the Uppsala countryside exploring for old runestones and ancient Viking mounds. Although the writer had left the University of Wisconsin to work with Professor Wold in Sweden, it turned out that Wisconsin under the leadership of Professor G. Box became the real center of time series analysis. It was the joint work of Box with Professor G. M. Jenkins [93] that actually brought the autoregressive (AR) process and the moving average (MA) process to the attention of the general scientific community. The brilliance of this work has made the names Box—Jenkins synonymous with time-series analysis. No achievement is better deserved. No person understands data better than Box in the application of statistical methods to obtain meaningful results.

In the 1960's Parzen [87] and Akaike [94] discussed autoregressive spectrum estimation, and this work led to their crucial work on autoregressive order-determining criteria [88] and [95]. Such criteria have made possible the widespread application of autoregressive spectrum estimation by researchers in diverse scientific fields. Akaike has provided a link between statistics and control theory with deep and significant results, and his work is of the highest tradition that science can provide. Young research workers can learn much by studying his writings well.

We wish we had more space and knowledge to expand upon this section, and those many statisticians whom we have not mentioned should remember that this history is by no means the final word. Someday we hope to write more fully on this subject, and we welcome all comments and suggestions.

Engineering use of spectral estimation

The purpose of this section is only to refer to the rest of the papers in this special issue of the *Proceedings of the IEEE*. These other papers cover the engineering use of spectral estimation much better than we could do here. There papers represent a living history of the present status of spectral estimation, and, in them and in the references which they give, the reader can find the works of the people who have made spectral analysis and estimation a vital scientific discipline today. As general references, we would especially like to mention the 1978 IEEE book edited by Childers [96], Haykin [97], the RADC Spectrum Estimation Workshop [98], the First IEEE ASSP Workshop on Spectral Estimation [99], and Ulrych and Bishop [100]. Although much progress has been made, much work yet remains to be done, and there is adventure for a research worker who sets his course in this rewarding and exciting field.

Acknowledgement

I want to express my sincere appreciation to Prof. D. R. Brillinger who let me freely use his paper, "Some history of data analysis of time series in the United States," in *History of Statistics in the United States*, edited by D. B. Owen and published by Marcel Dekker in 1976. I want to thank Dr. J. Makhoul for sending me notes on maximum likelihood and lattice networks. I want to especially thank the authors cited in the Reference Section whose constructive comments materially improved this paper. In writing an historical paper, we should include a thousand references instead of one hundred, so important statistical contributions have unfortunately been left out. Of course, we have purposely not included engineering contributions (as they are covered in the rest of this special issue) but there is never a clear cut line, and so in this sense other important work has also been left out. However, all such omissions are not intentional, and we will gladly try to rectify any situation in some appropriate future publication. Finally, most of all, I want to thank Prof. J. Tukey for the support and help he gave me on spectrum estimation thirty years ago at MIT for which I am forever grateful.

References

[1] I. Newton, Optics, London, England, 1704.

[2] B. Taylor, Methodus Incrementorum Directa et Inverse. London, England, 1715.

[3] D. Bernoulli, Hydrodynamics Basel, Switzerland, 1738.

[4] L. Euler, Institutiones Calculi Differentialis. St. Petersburg, Russia, 1755.

[5] J. L. Lagrange, V4orie des Fonctions Analytiques Paris, France, 1759.

(6) J. Fourier, T14orie Analytique de la Chaleur. Paris, France: Didot, 1822.

[7] C. Sturm, "Memoire sur les equations differentielles lineaires du second ordre," Journal de Mathimatiques Pures et Appliquies, Paris, France, Series 1, vol. 1, pp. 106-186, 1836.

[8] J. Liouville, "Premier memoire sur la theorie des equations differentielles linearies et sur le developpement des fonctions en series," Journal de Mathi'matiques Pures et Appliqi4es, Paris, France, Series 1, vol. 3, pp. 561-614, 1838.

[9] G. Green, Essay on the Application of Mathematical Analysis to the Theories of Electricity and Magnetism. Nottingham, En-land, 1828.

[10] E. Schrodinger, Collected Papers on Wave Mechanics London, England: Blackie, 1928.

[11] W. Heisenberg, The Physical Principles of Quantum Theory. Chicago, IL: University of Chicago Press, 1930.

[12] J. von Neumann, "Eigenwerttheorie Hermitescher Funcktional-operatoren," Math. Ann., vol. 102, p. 49, 1929.

[13] —,Mathematische Grundlagen der Quantenmechanik. Berlin, Germany: Springer, 1932.

[14] M. von Smoluchowski, The Kinetic Theory of Matter and Electricity. Leipzig and Berlin, Germany, 1914.

[15] A. Einstein, "On the theory of the Brownian movement," Annalen der Physik, vol. 19, pp. 371-381, 1906.

[16] N. Wiener, "Differential space," J. Math. Phys., vol. 2, p. 131.

[17] A. Schuster," On the investigation of hidden periodicities with application to a supposed 26-day period of meterological phenomena," Tern Magnet., vol. 3, pp. 13-41, 1898.

[18] G. U. Yule, "On a method of investigating periodicities in disturbed series, with special reference to Wolfer's sunspot numbers," Phil Trans Roy. Soc. London, A, vol. 226, pp. 267-298.

[19] Robinson, E. A., Predictive Decomposition of Time Series with Applications to Seismic Exploration, 265 pp., Ph.D. Thesis, M.I.T., 1954. Reprinted as MIT GAG Report No. 7, 1954. Reprinted in Geophysics, vol. 32, pp. 418-484, 1967.

[20] —, An Introduction to Infinitely Many Variates London, England: Griffin, 1959, p. 109.

[21] N. Wiener, "Generalized harmonic analysis," Acta Math., vol. 55, pp. 117-258, 1930.

[22] A. Y. Khintchine, "Korrelations theorie der Stationiren Stochastischen Prozesse," Math. Ann., vol. 109, p. 604.

[23] N. Wiener, Cybernetics Cambridge, MA: MIT Press, 1948.

[24] —, Extrapolation, Interpolation, and Smoothing of Stationary Time Series with Engineering Applications, MIT NDRC Report, 1942, Reprinted, MIT Press, 1949.

[25] —, The Fourier IntegraL London, England: Cambridge, 1933.

[26] P. Dirac, Principles of Quantum Mechanics New York: Oxford University Press, 1930.

[27] O. Heaviside, Electrical Papers, vol. I and II. New York: Macmillan, 1892.

[28] J. von Neumann, "Uber Funktionen von Funktional opera-toren, "Ann. Math., vol. 32, p. 191.

[29] N. Wiener, I Am a Mathematician. Cambridge, MA: MIT Press, 1956.

[30] N. Levinson, "A heuristic exposition of Wiener's mathematical theory of prediction and filtering," Journal of Math. and Physics, vol. 26, pp. 110-119, 1947.

[31] —, The Wiener RMS (root mean square) error criterion in filter design and prediction, J. Math. Phys., vol. 25, pp. 261-278, 1947.

[32] G. P. Wadsworth, Short-Range and Extended Forecasting by Statistical Methods, Air Weather Service, Washington, DC, 1948.

[33] H. R. Seiwell, "The principles of time series analyses applied to ocean wave data," in Proc. Nat. Acad Sci., U.S.., vol. 35, pp. 518-528, 1949.

[34] J. W. Tukey, "The sampling theory of power spectrum estimates," in Proc. Symp. AppL Autocorr. Anal: Phys Prob. U.S. Off Naval Res (NAVEXOS-P-725), 1949. Reprinted in J. Cycle Res., vol. 6, pp. 31-52, 1957.

[35] G. P. Wadsworth, E. A. Robinson, J. G. Bryan, and P. M. Hurley, "Detection of reflections on seismic records by linear operators," Geophys, vol. 18, pp. 539-586, 1953.

[36) J. W. Tukey and R. W. Hamming, "Measuring noise color 1," Bell Lab. Memo, 1949.

[37] J. W. Tukey, Measuring Noise Color, Unpublished manuscript prepared for distribution at the Institute of Radio Engineers Meeting, Nov. 1951.

[38] S. M. Simpson, Time Series Computations in FORTRAN and FAP Reading, MA: Addison-Wesley, 1966.

[39] N. Wiener, Bull. Amer. Math. Soc., vol. 72, pp. 1-145.

[40] N. Wiener and P. Masani, "The prediction theory of multivariate stochastic processes," Acta Math., vol. 98, pp. 111-150, 1957.

[41] —, "On bivariate stationary processes," Theory Prob. AppL, vol. 4, pp. 300-308.

[42j R. A. Wiggins and E. A. Robinson, "Recursive solution to the multichannel filtering problem," J. of Geophys Res., vol. 70, pp. 1885-1891, 1965.

[43] E. A. Robinson, Statistical Communication and Detection with Special Reference to Digital Data Processing of Radar and Seismic Signals. London, England: Griffin, 1967. Reprinted with new title, Physical Applications of Stationary Time Series. New York: Macmillan, 1980.

[44] J. W. Tukey, "An introduction to the calculations of numerical spectrum analysis," in Spectral Analysis of Time Series, B. Harris, Ed. New York: Wiley, 1967, pp. 25-46.

[45] H. Press and J. W. Tukey, "Power spectral methods of analysis and their application to problems in airplane dynamics," Bell Syst Monogr., vol. 2606, 1956.

[46] R. B. Blackman and J. W. Tukey, "The measurement of power spectra from the point of view of communications engineering," Bell Syst Tech. J., vol. 33, pp. 185-282, 485-569, 1958; also New York: Dover, 1959.

[47] J. W. Tukey, "The estimation of power spectra and related quantities," On Numerical Approximation, R. E. Langer, Ed. Madison, WI: University of Wisconsin Press, 1959, pp. 389-411.

[48] —, "An introduction to the measurement of spectra," in Probability and Statistics, U. Grenander, Ed. New York: Wiley, 1959, pp. 300-330.

[49] —, "Equalization and pulse shaping techniques applied to the determination of the initial sense of Rayleigh waves," in The Need of Fundamental Research in Seismology, Appendix 9, Department of State, Washington, DC, 1959, pp. 60-129.

[50] B. P. Bogert, M. J. Healy, and J. W. Tukey, "The frequency analysis of time series for echoes; cepstrum pseud-autocovariance, cross-cepstrum and shape-cracking," in Time Series Analysis, M. Rosenblatt, Ed. New York: Wiley, 1963, pp. 201-243.

[51] J. W. Tukey, "Discussion emphasizing the connection between analysis of variance and spectrum analysis," Technometrics, vol. 3, pp. 1-29, 1961.

[52] —, "The future of data analysis," Ann. Math. Statist., vol. 33, pp. 1-67, 1963.

[53] —, "What can data analysis and statistics offer today?" in Ocean Wave Spectra, Nat. Acad. Sci., Washington, DC, and Prentice-Hall, Englewood Cliffs, NJ, 1963.

[54] —, "Uses of numerical spectrum analysis in geophysics," Bull Int Statist. Inst., vol. 39, pp. 267-307, 1965.

[55] —, "Data analysis and the frontiers of geophysics," Science, vol. 148, pp. 1283-1289, 1965.

[56] W. J. Pierson and L. J. Tick, "Stationary random processes in meteorology and oceanography," Bull. Int Statist. Inst., vol. 35, pp. 271-281, 1957.

[57] N. R. Goodman, "On the joint estimation of the spectra, cospectrum and quadrature spectrum of a two-dimensional stationary Gaussian process," Science Paper no. 10, Engineering Statistics Laboratory, New York University, New York, 1957.

[58] W. H. Munk and G. J. F. MacDonald, The Rotation of the Earth. New York: Cambridge University Press, 1960.

[59] M. J. Healy and B. P. Bogert, "FORTRAN subroutines for time series analysis," Commun. Soc. Computing Machines, vol. 6, pp. 32-34, 1963.

[60] E. C. Bullard, F. E. Ogelbay, W. H. Munk, and G. R. Miller, A User's Guide to BOMM, Institute of Geophysics and Planetary Physics, University of California Press, San Diego, 1966.

[61] E. Parzen, Time Series Analysis Papers. San Francisco, CA: Holden-Day, 1967.

[62] E. A. Robinson, Multichannel Time Series Analysis with Digital Computer Programs San Francisco, CA: Holden-Day, 1967.

[63] C. W. J. Granger, Spectral Analysis of Economic Time Series. Princeton, NJ: Princeton University Press, 1964.

[64] J. Neyman and E. L. Scott, "Statistical approach to problems of cosmology," J. Roy. Statist. Soc., Series B, vol. 20, pp. 1-43, 1958.

[65] U. Grenander and M. Rosenblatt, Statistical Analysis of Stationary Time Series New York: Wiley, 1957.

[66] N. Wiener, Nonlinear Problems in Random Theory. Cambridge, MA: MIT Press, 1958.

[67] —, "Rhythm in physiology with particular reference to encephalography," Proc. Roy. Virchow Med. Soc., NY, vol. 16, pp. 109-124, 1957.

[68] J. W. Cooley and J. W. Tukey," An algorithm for the machine calculation of Fourier series," Math. Comput., vol. 19, pp. 297-301, 1965.

[69] J. W. Cooley, P. A. W. Lewis, and P. D. Welch, "Historical notes on the fast Fourier transform," IEEE Trans Audio Electro-acoust., vol. AU-15, pp. 76-79, 1967.

[70] C. Bingham, M. D. Godfrey, and J. W. Tukey, "Modern techniques of power spectrum estimation," IEEE Trans. Audio Electroacoust., vol. AU-15, pp. 56-66, 1967.

[71] E. O. Brigham and R. E. Morrow, "The fast Fourier transform," IEEE Spectrum, pp. 63-70, 1967.

[72] IEEE Trans. Audio Electroacoust (Special issue on Fourier transform), B. P. Bogert and F. Van Veen, Eds., vol. AU-15, June 1967 and vol. AU-17, June 1969.

[73] D. R. Brillinger, Time Series: Data Analysis and Theory. New York: Holt, Rinehart, and Winston, Inc., 1974; revised edition by Holden-Day, San Francisco, CA, 1980.

[74] D. R. Brillinger and M. Rosenblatt, "Computation and interpretation of k-th order spectra," in Spectral Analysis of Time Series, B. Harris, Ed. New York: Wiley, 1967, pp. 189-232.

[75] J. Capon, "Applications of detection and estimation theory to large array seismology," Proc. IEEE, vol. 58, pp. 760-770, 1970.

[76] Stochastic Point Processes, P. A. W. Lewis, Ed. New York: Wiley, 1972.

[77] P. A. W. Lewis, A. M. Katcher, and A. H. Weiss, "SASE IV-An improved program for the statistical analysis of series of events," IBM Res Resp., RC2365, 1969.

[78] Proc. Symp. Walsh Functions, 1970-1973.

[79] A. Cohen and R. H. Jones, "Regression on a random field," J. Amer. Statist. Ass., vol. 64, pp. 1172-1182, 1969.

[80] E. A. Robinson, "Predictive decomposition of seismic traces," Geophysics, vol. 22, pp. 767-778, 1957

[81] E. A. Robinson and H. Wold, Minimum-Delay Structure of Least-Squares and Eo Ipso Predicting Systems, Time Series Analysis, (edited by M. Rosenblatt), John Wiley, N.Y., pp. 192-196, 1963.

[82] W. A. Schneider, Integral formulation for migration in two and there dimensions. Geophysics, vol. 43, pp. 49-76, 1978.

[83] M. T. Silvia and E. A. Robinson, Deconvolution of Geophysical Time Series in the Exploration of Oil and Natural Gas Amsterdam, The Netherlands: Elsevier, 1979.

[84] F. Itakura and S. Saito, "Analysis synthesis telephony based on on the maximum-likelihood method," in Proc. Sixth Int. Conf. on Acoustics, (Tokyo, Japan), 1968.

[85] E. Parzen, "Modern empirical spectral analysis," Tech. Report N-12, Texas A&M Research Foundation, College Station, TX, 1980.

[86] J. Makhoul, "Stable and efficient lattice methods for linear prediction," IEEE Trans Acoust., Speech, Signal Processing, vol. ASSP-25, pp. 423-428, 1977.

[87] E. Parzen, "An approach to empirical time series," J. Res Nat. Bur. Stand., vol. 68B, pp. 937-951, 1964.

[88] —, "Some recent advances in time series modeling," IEEE Trans Automat Contr., vol. AC-19, pp. 723-730, 1974.

[89] M. Priestley, Spectral Analysis and Time Series, 2 volumes, London, England: Academic Press, 1981.

[90] H. Wold, A Study in the Analysis of Stationary Time Series. Stockholm, Sweden: Stockholm University, 1938.

[91] P. Whittle, Hypothesis Testing in Time-Series Analysis. Uppsala. Sweden, Almqvist and Wiksells, 1951.

[92] —, The statistical analysis of a seiche record, J. Marine Res., vol. 13, pp. 76-100, 1954.

[93] G. Box and G. M. Jenkins, Time Series Analysis, Forecasting, and ControL Oakland, CA: Holden-Day, 1970.

[94] H. Akaike, `Power spectrum estimation through autoregressive model fitting," Ann. Inst. Statist. Math., vol. 21, pp. 407-419, 1969.

[95] —, "A new look at statistical model identification," IEEE Trans. Autom. Contr., vol. AC-19, pp. 716-723, 1974.

[96) D. G. Childers, Modern Spectrum Analysis. New York: IEEE Press, 1978.

[97] S. Haykin, Nonlinear Methods of Spectral Analysis Berlin, Germany: Springer, 1979.

[98] Proc. of the RADC Spectrum Estimation Workshop, Rome Air Development Center, Griffiss Air Force Base, NY, 1979.

[99] Proc. of the First ASSP Workshop on Spectral Estimation, McMaster University, Hamilton, Ontario, Canada, 1981.

[100] T. J. Ulrych and T. N. Bishop, "Maximum entropy spectral analysis and autoregressive decomposition," Rev. Geophys., vol. 13, pp. 183-200, 1975.

[101] P. R. Gutowski, E. A. Robinson, and S. Treitel, "Spectral estimation: Fact or fiction," IEEE Trans Geosci. Electronics, vol. GE-16, pp. 80-84, 1978. Reprinted in D. G. Childers, Modern Spectrum Analysis New York: IEEE Press, 1978.

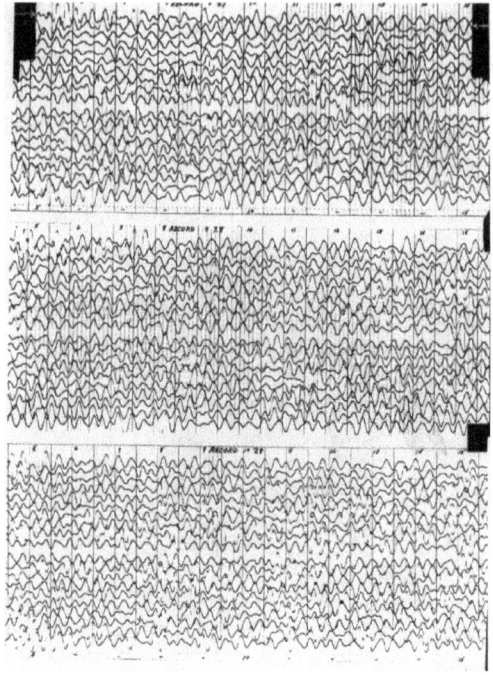

Some of the analog seismic signals (top) that were digitized and digitally processed on the Whirlwind computer (bottom) in 1952 and 1953

www.ingramcontent.com/pod-product-compliance
Lightning Source LLC
Chambersburg PA
CBHW051644170526
45167CB00001B/328